PENGUIN BOOKS

GORGON

Peter Ward, a recognized authority on mass extinctions, is professor of geological sciences at the University of Washington in Seattle. His books include *Future Evolution*, *The End of Evolution*, and with Donald Brownlee, *Rare Earth* and *The Life and Death of Planet Earth*. He lives in Seattle, Washington.

A gorgonopsian by Alexis Rockman, 9" x 12", acrylic and Permian sediment on paper. Collection of Dorothy Spears.

GORGON

The Monsters That Ruled the Planet
Before Dinosaurs and How They Died in
the Greatest Catastrophe in Earth's History

Peter D. Ward

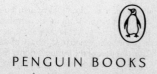

PENGUIN BOOKS

PENGUIN BOOKS
Published by the Penguin Group
Penguin Group (USA) Inc., 375 Hudson Street, New York, New York 10014, U.S.A.
Penguin Group (Canada), 10 Alcorn Avenue, Toronto,
    Ontario, Canada M4V 3B2 (a division of Pearson Penguin Canada Inc.)
Penguin Books Ltd, 80 Strand, London WC2R 0RL, England
Penguin Ireland, 25 St Stephen's Green, Dublin 2, Ireland (a division of Penguin Books Ltd)
Penguin Group (Australia), 250 Camberwell Road, Camberwell,
    Victoria 3124, Australia (a division of Pearson Australia Group Pty Ltd)
Penguin Books India Pvt Ltd, 11 Community Centre,
    Panchsheel Park, New Delhi – 110 017, India
Penguin Group (NZ), cnr Airborne and Rosedale Roads, Albany,
    Auckland 1310, New Zealand (a division of Pearson New Zealand Ltd)
Penguin Books (South Africa) (Pty) Ltd, 24 Sturdee Avenue,
    Rosebank, Johannesburg 2196, South Africa

Penguin Books Ltd, Registered Offices: 80 Strand, London WC2R 0RL, England

First published in the United States of America by Viking Penguin,
a member of Penguin Group (USA) Inc. 2004
Published in Penguin Books 2005

10 9 8 7 6 5 4 3 2 1

Photographs by the author.

Frontispiece: A gorgonopsian by Alexis Rockman, 9" x 12", acrylic and
Permian sediment on paper. Collection of Dorothy Spears.

THE LIBRARY OF CONGRESS HAS CATALOGED
THE HARDCOVER EDITION AS FOLLOWS:
Ward, Peter Douglas.
    Gorgon : paleontology, obsession, and the greatest
catastrophe in earth's history / Peter Ward.
        p.   cm.
    Includes bibliographical references and index.
    ISBN 0-670-03094-5 (hc.)
    ISBN 0 14 30.3471 5 (pbk.)
    1. Catastrophes (Geology).   2. Geology, Stratigraphic—Permian.   3. Geology, Structural—
South Africa—Karoo.   I. Title.
    QE674.W37 2004
    576.8'4—dc21   2003047953

Printed in the United States of America

For Christine

Preface

Near the southern margin of the African continent, a ringwall of mountains holds up a vast interior desert known as the Great Karoo. It is far from anywhere, in both space and time. The Karoo is a Lost World.

In the Khoikhoi language, Karoo means "land of thirst." It is a broad plateau that is either too hot or too cold for human comfort and always too dry; a place where water is a rarity and where your skin dries and ages even at night. The Karoo is a region where it can snow in summer yet where the temperature can reach more than a hundred degrees Fahrenheit on the shortest day of winter. On some days the Karoo will require you to drink four liters of water, and even then you will never urinate, so rapidly does the furnacelike heat suck moisture from your body; yet the very next day the Karoo can deliver the glacial cold that is but a premonition of death's long future sovereignty.

It is a land where incessant wind coats all with a new daily layer of dust, where each craggy rock in a land filled with them may harbor a Cape cobra, or khals, or puff adder, the venomous serpents of the Karoo that will strike without warning and leave your flesh rotting for months afterward, assuming that you can get to an antivenin serum in time to survive the bite. The Karoo is a place where scorpions and poisonous spiders abound, where a thousand varieties of flies seek to feed on you, loving your eyes most of all, where ticks with pathogenic viruses cling to every bush poised to drop onto any passing mammal and can be always be found on your tent floor at night, using your heat and carbon-dioxide signature as a means to find you and bore their infected heads into your skin and suck your blood. It is a place where mosquitoes swarm after each infrequent, torrential rain, buzzing around your head in the night, their faint but insidious, high-pitched whine ensuring that you will never sleep.

It is a place immense and flat, filled with low shrubs and scrub festooned with thorns that will break off and fester in your skin, releasing their alkaloid toxins. It is a landscape once home to a flourishing variety of larger antelope and other African game but now mainly host to huge herds of sheep all striving to find any greenery on disappearing small farms, a land where the few remaining predatory cats are hunted mercilessly and any animal on scattered game parks can be shot and stuffed for a price. It is a land where bands of black human rovers will murder white farmers for food or simply for hate, a place where many of the white settlers who own or control all of the region's wealth live with a siege mentality and travel armed and ready for trouble, a place where the few last holdouts of the vicious apartheid system still harbor long grudges and a deep-seated and obdurate racism.

Why would you leave a loving family and a comfortable house to spend weeks or months in this lonely place, to live in a small, cold tent and sleep on the unyielding ground, your mouth gritty with the deepening dust brought by the howling, never-ending wind, where for a toilet you squat each morning in the dusty veld like our earliest human ancestors of a million years ago? Why would you go to a place where the drinking water is malodorous and diseased, where a bath is a little basin and a cloth or a hoseful of near freezing sulfurous water from an irrigation ditch, where the diet is mainly meat and vegetables are virtually nonexistent, where your face turns red and cracked from the sun and your body rapidly becomes covered with infected sores from the thorns and cuts on the barbed wire and the falls onto sharp rock as you work twelve-hour days in the heat and cold looking for elusive treasures that are often never found at all, a place where the nearest telephone is often a half day's drive away and the only news is of the price of wool and the list of hate crimes? Who would willingly do this? Who would come to such a place?

You would arrive in a brutal land, ancient and rocky, where the history is long and cataclysmic, the lives short and the deaths violent. It is a land of hardship and poverty, of wars near and vicious. But it also is domed by a sky of deepest blue overriding palettes of rainbow desert colors. Endless horizons and lofty stone ramparts strain into this vaulted African sky, a sky filled at night with the splendors of stars unknown and unseen in the far north, where the pale eyes of the Magellanic Clouds, great nearby sister galaxies visible only in the Southern Hemisphere, vie for eminence with the Southern Cross in brazen glory, punctuating a blackness so deep that its clutched stars seem touchable. The fleet springbok, eland, reebok, zebra, and the more lumbering gemsbok, bontebok, wildebeest, and kudu can still be found in the wilder corners of this land, where aardvarks destroy the towering termite nests at night. Here you can see the glorious secretary birds do their snake-charming dances amid the thorny flats, while gigantic, magnificent blue cranes fly overhead each morning and evening like great soaring monoplanes to land stiffly in the bristly veld in their search for food. A hundred colored songbirds will throng the prickly bracken and serenade you with their tunes, giant land tortoises will examine you with indignation, and troops of monkeys and baboons will thrash the bush with boisterous vitality and chatter at you with ribald good humor as you do your arcane human business in their huge and wild land.

You are in a land of sweet smells and clean air, of heat lightning in the distance, sharing camp with companions both mortally frail and strong beyond belief in a country where freedom has only been newly won and still tastes like the sweetest springwater. Many of the farmers here will gather you in for a drink, a cup of tea, a chat; they will help you to understand what life on a frontier is like and always make you feel that their home is yours as well. You have arrived in a place where you will find fellow travelers in this life's journey who have come, like you, to seek answers to a great mystery; colleagues who share the heat and cold and discomfort and dirt without complaint; men and women who can cook the most delicious dinners of freshly killed meat mixed with wild garlic, onions, and desert spices over the open wood fires; friends who watch for your safety on the craggy slopes and more dangerous rocky defiles, who greet each morning with cheer and work tirelessly through the long day, who live the same hard life and in some way ease the harshness through this sharing.

You have come to a land so large that you can finally no longer escape from yourself through the telephone calls and e-mails that have become so addictive and time-devouring and usually trivial, a place where the mindless televisions and the hectic pace of the industrialized world have been left behind. You have come to a place where you will have hours with only yourself for company, where the unending sky and baked rocks will sooner or later become a mirror on all sides that force you to see yourself for what you really are, have been, will be—a vision that may not lead to change but must, finally, end in acceptance and, in ways larger or smaller, in peace.

Why would you come? For some it is to search for truth, for discovery, for knowledge, to find fossil treasures, the bones of long-distant ancestors, spectral skulls leering but silent that are among the most precious of all the earth's fossil record, antiquities of immense beauty and rarity that when found cause the heart to race and bring cries of joy whooping among the rocks as all age falls away and the best moments of childhood are experienced once more. For some it is to unearth the slender clues to one of science's most enduring and important problems, for the Karoo must hold answers to why more than 90 percent of living organisms suddenly died off 250 million years ago in the greatest catastrophe ever to have assaulted life on this planet—a catastrophe so great that it reset the evolutionary clock, utterly changing the river course that is the history of life. It was an event that brought down one set of animals and gave the world to another, that perhaps set back the Age of Mammals by nearly 200 million years, that made the death of the dinosaurs some 65 million years ago pale in comparison. A catastrophe the likes of which could once again threaten life on this planet and may indeed be doing so now. Your work here may not answer any of these questions; they may be intractable. But they will stay locked in secrecy unless you try.

You are a soldier of low rank; you live and campaign in the field. Ignorance is the enemy, curiosity the weapon of choice. Each morning you don the dusty jeans, the comfortable boots and cool cotton shirt. You strap on belts, attach the aged leather of field case, Brunton compass, GPS, and loupe holder; gather the cold steel of hammer and chisel, the comfortable hat, and await the day's orders. A particular river course to be searched. A cliff site that needs to be scrutinized. What fossils to search for, how the bone will look, why this particular day's work is important. You regard the other gathered faces of the team that you will be with today and smile. There is nowhere you would rather be.

Why would you come?

Who wouldn't?

Acknowledgments

A book covering events of more than a decade owes debts to many people, debts that can never be paid in full. I would like to thank the many friends in South Africa who allowed me to work and live there: the Rubidge family, James Kitching, Mike Cluver, the Karoo Paleontology Unit at the South African Museum, including Georgina Skinner, Paul Smith, Hedi Stummer, and of course the star of this book, Roger Smith, and his heroic wife Sally. Thanks to my research partner Joe Kirschvink, and Ken MacLeod, Roger Buick, Greg Retallack, and Diane Richards for reading a previous draft and helping me see what writing is. And thanks to my friend Alexis Rockman for sharing so much of this. Much thanks to the funding agencies that allowed me to work there (the National Science Foundation and the NASA Astrobiology Institute). Thanks as well to my family: Chris, Nicholas, and Patrick. At Viking Penguin my editor, Sarah Manges, made the changes that mattered and fought to lift this story out of the tired Science Trade Book genre.

Imagine: human scientists!

Contents

Introduction

The farmhouse has thick white stone walls, faded green trim, and an old metal roof now partially unfastened and moaning in each gust of dry African wind. It is Dutch Colonial in style, with a large living room and blackened fireplace, yellow wood floors, and high ceilings. There is no central heating or indoor plumbing. Wildlife of moderate diversity now resides in the upper story, most of it winged but with the odd resident quadruped in there as well, judging from the scurrying sounds and rodent droppings scattered about. Layered, dusty spiderwebs are the final touch of decoration, creating a tent city that gently pulses from the breezes entering the house through the many broken windows.

The front of the house is graced by a handsome stone lintel, simply inscribed: TWEEFONTEIN, APRIL 30, 1863. A wide front porch beneath this dedication is lined with inset flagstones, many scored with graffiti. These now-ancient inscriptions are in square, childlike script, not at all the sort of writing polluting a modern building. Beautifully carved initials: P.F., 22 NOV. 1897. On another, P.F. appears again, JAN. 25, 1898. There is a C.L., M.F., C.L., 26/6/1901. And a new name, JOHN GOED, 1903. Then nothing more.

The yard in front of the house is shifting dirt and dust, now home to stunted thornbushes native to the surrounding Karoo Desert. Termite mounds give relief to what was once a front lawn.

Dead trees surround the house, obviously planted long ago for shade, but now casting the anorexic shadows of skeletons. Other trees farther from the house have fallen and are being reduced to dirt by the African decomposers.

To the north of the house, a barn has also fallen inward, and beyond that an obviously older building has decayed into ruin as well, this one made of individual stones, carefully and methodically fashioned together to form what once must have been perfectly squared cobble walls, a legacy of craftsmen with time on their hands. A stone fence of the same exquisite craftsmanship runs parallel to the older building and then disappears into a pile of rubble.

The front of the house looks west, into the sunsets, while the back is overlooked by the giant stone rampart known as Lootsberg Pass, a wall of rock that rises a thousand feet above the farm. A hard walk of ten minutes from the back of the farm takes one onto the slope that rises up onto the beginning of this wall of high sedimentary rock and capping volcanic layers. At the start of this slope, between the house and the long mountain range behind, is one final testament to a former human existence here: the family graveyard.

The graveyard is perhaps twenty-five feet on a side, and to my eye it looks as if it were constructed for a Hollywood haunted-house B movie, except that this is the real thing. The graves are enclosed by a low stone wall topped by Victorian black grillework of densely packed and upthrusting iron points. Desiccated weeds now fill the walled graveyard, partially obscuring three large marble headstones. Legions of ants march between the headstones in military formation, skirting the brown wrack of dead bracken on their mindless journey through this Boer cemetery. The sun shines brightly on the dead this day, as it does on most days in the Karoo; a squadron of flies patrols its perimeter. Wind

blows on the weeds among the headstones, and an African equivalent of a tumbleweed rolls up to the wall surrounding the
graves, becoming suspended there. Sheep dung lies everywhere
around, the universal currency of the Karoo.

The largest headstone is inscribed with two names: Anna
Catherine Jacobus Fouche, born 1831, died 1892, and her husband, Francouis Petrus Fouche, born 1825, died 1894. The headstone is still pristine white save for ancient fingers of brown
lichen, the marble imported—surely not of African origin—and
beautifully carved. A second headstone, smaller but of the same
marble, lies several feet to the side of the first. Carel Hercules
Fouche, born 1856, died 1898. The third headstone, the same size
as the second and the same quality as both, proclaims the resting
place of Jacobus Albertus Fouche, born 1859, died 1900. And next
to these three graves, all in a row, lies a fourth. An extended pile
of stones shows another burial, but there is no headstone, no indication of who—or what—lies here. A family servant? A favorite dog? Or does this grave hold another Fouche whose
headstone has been stolen or who did not merit one in punish-
ment for sins long forgotten?

There is plenty of space for more bodies here. Whoever designed this graveyard had a bigger or longer lasting family in
mind, and, based on the scratching on the farmhouse stones, it
seems that at least some Fouches survived the rather surprising
deaths of the rest of the Fouche family, for the dates of death on
the headstones suggests a rather sudden end for four members of
this family. How often does a family, or perhaps most of one, die
out in such a short interval of time? Anna Catherine was the first
to go, in 1892, at age sixty-one; her husband, then aged sixty-
nine, followed two years later—a pattern often seen among
grieving spouses. But the two sons, Carel and Jacobus, lasted not
much longer: Carel died five years after his father, at the age of

forty-two, and his younger brother, Jacobus, died in 1900 at the young age of forty-one. In eight years they were all gone. The headstones give no clue to the cause of this family's demise. Four members of a family so quickly changed from living to dead. Why? There is also no clue to be had in the nearest register of information, the Boer records in the public library in the nearby town of Graaff-Reinet. The Fouches are listed, their births and deaths recorded, but no cause of death is given. And who survived them? Who is the P.F. who scratched his name in the stone front of the house after Jacobus Fouche died? This family died out, or nearly did so, a century ago. All record of how this tragedy occurred is now lost; we have but the bodies and the birth and death dates.

The irony is sublime. Quite unknowingly, the Fouches placed their cemetery on a far older graveyard, a cemetery planted 250 million years ago whose headstone is the high rampart of rock we now call Lootsberg Pass, a sandstone and shale monument soaring up over the gothic iron grillework and marble of the Fouches' last resting place. This older fossil grave site marks the death of as many as 90 percent of all species that lived in the world at that time. But what hope is there to understand a quarter-billion-year-old tragedy, when a century-old series of deaths becomes so quickly opaque to our comprehension? At the same time, how can we *not* try to understand death, both recent and remote? Especially if those deaths could hold clues to the future as well as the past?

This book is about two things. The first is the search for the cause of the long-ago mass extinction so clearly marked in Africa and elsewhere around the globe, a calamity devastating enough to bring life on the earth into its most severe near-death experience. If we can find the cause, then perhaps we can predict whether such a calamitous event might be in our future as well as our planet's past. To that end I will tell a history of how research about this question unfolded in Africa. The second topic of the pages to come, however, is along a quite different tangent: why scientists would so doggedly, and at the cost of such hardship, devote their lives to pursuing this ancient killer. The popular view is that paleontology is a romantic profession. While that may be, after years of work in the heat and cold of a high desert, romance wears thin. And yet, as we shall see, the scientists portrayed in this saga slogged on and on, year after year, for little pay and even less glory. But first the extinction.

▪ ▪ ▪

Long ago, at the beginning of Paleozoic time (some 540 million years ago), life flourished in the seas, but only bacteria and perhaps some low fungi and moss lived on land. This state of affairs did not last, and by 300 million years ago, a flourishing diversity of creatures, including a rich suite of vertebrates, inhabited the continents. By 260 million years ago land animals were as common as dinosaurs and mammals would ever be. Some would have seemed familiar enough: smaller lizards and salamander-like amphibians that were quite similar in shape (if not ancestry) to animals still living today. But the larger animals would have been totally unfamiliar, for they belonged to an assemblage of land animals completely unknown to popular culture—the mammal-like reptiles. There were many varieties of them: herbivores, scavengers, diggers, tree dwellers, and hunters.

The largest land-dwelling carnivore of 260 million years ago, the *T. rex* of its day, was about the size of a modern lion. But other than having a similarity in size, these ancient carnivores were surely most unlionlike in most other traits. They would have had huge heads with very large, saberlike teeth, large lizard eyes, no visible ears, and perhaps a mixture of reptilian scales and tufts of mammalian hair. Their tails would have been relatively long and reptilian, and they would have moved about on rather stumpy legs with broad, flat feet bearing hideously large claws, not retractable like those of cats, but more like those of a wolf. They must have loped after prey in a relentless fashion, legs splayed out slightly to the side, outsize, fanged heads held high. They are monsters from the dark recesses of our imagination; they are the stuff of nightmares, half dragon, half lion, cunning land sharks adapted for eating and killing. They fed on every type of larger animal then existing, some of which were the direct ancestors of all mammals living today (including humanity). They are called Gorgons.

The Gorgons ruled a world of animals that were but one short evolutionary step away from being mammals and that evolved into true mammals long before there was any Age of Dinosaurs. The Age of Mammals was set to begin some 250 million years ago but was summarily ended, or at least vastly delayed.

The very destructiveness of this mass-extinction event has led to new theories about the importance of catastrophe in determining the nature of the makeup of life on earth. To many scientists it suggests that we live in a very troubled and dangerous world, one molded and sculpted by a succession of calamities, and that these catastrophes, called mass extinctions, are the largest-scale evolutionary events affecting the biology of our planet.

There can be no doubt that mass-extinction events—there have been five, in the last 500 million years, that killed off more than half of species on earth—have led to wholesale biological changes. After all, the most famous of the mass extinctions, which killed off all dinosaurs 65 million years ago, was immediately followed by an entirely new mix of species that we call the Age of Mammals. Yet there might be an even more profound implication. Paradoxically, the mass extinctions might be of great importance in understanding the nature of the cosmos—events that might tell us something about the frequency of life in the galaxy and where to look for it. That, too, is part of the story to come.

And finally the *why:* Why dedicate a career and life to such a pursuit? That story is more complex, harder told, but perhaps the more interesting of the two intertwined tales I have tried to write in this book. You be the judge.

A graphical representation of our decade of work in the Karoo. This diagram, being published in the scientific literature, shows how the ranges of the mammal-like reptiles (from the collecting trips of Roger Smith and myself) match up with magnetostratigraphy (the work of Joe Kirschvink) and the isotope work, which began with Ken MacLeod and was expanded by my efforts. This graph shows that there was a rapid extinction at the same time as a major change in isotope values, indicating a plant die off. However, the data here do not support the type of extinction ending the Cretaceous, brought about by meteor impact. This seems to be due to sudden global warming and change in the atmosphere.

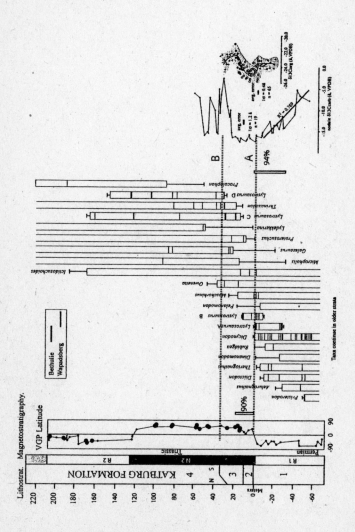

GORGON

Chapter 1

Arriving

SEPTEMBER 1991

The ends of long love affairs are never painless.

For nearly a decade, I had been hopelessly, shamelessly in love with the adventure of solving a wonderful scientific mystery: what killed the dinosaurs and their ilk, 65 million years ago, in what had been named the Cretaceous-Tertiary, or K/T mass extinction. Since 1980 my world—the world of geologists, astronomers, paleontologists, geochemists, in sum, those of us interested in earth history—had been involved in the affair of our lives: the mystery behind the death of the aforesaid dinosaurs. It was a consuming affair, all-important, satisfying, demanding, engaging, frustrating, emotional, crazed. It took us to faraway places, prompted us to spend fortunes and youth, to give allegiance, to become involved in public debate and controversy and censure and comradeship on epic scales (or as epic as things get for a scientist anyway). It involved the mobilization of an army of scientists, of gatherings large and small for the dissemination of knowledge. There were riches to be gained in prestige and sometimes money, the not inconsiderable stakes of careers raised up or torn down among the various emotions of anger, nobility, jealousy, altruism, and unrequited passion by all involved. But, most important, the affair lent a wonderful *scratch* to that constant itch of curiosity that drives science and its motley crew of practitioners. *What killed the dinosaur and so much else?* Was it slow climate change, or pulsing volcanoes, or even competition from egg-eating mammals as scientists had thought ever since the first dinosaurs were discovered now nearly two centuries ago? Or was it something much more dramatic—like a really big rock falling from space? It was this latter assertion, arriving as a clarion call to action in 1980 with the now famous publication by the Alvarez group from Berkeley, that galvanized a scientific community into action. Those of us called fanned out, did our work, and within a decade ultimately proved that a ten-kilometer asteroid crashing into the earth caused an enormous and catastrophic mass extinction.

By the fall of 1991, after more than a decade of research and discovery, the affair was over. The controversy ended. Data and falsification laid bare the lie of a long, slow event caused by gradual climate change or egg-eating mammals or whatever. But resolution took away our lovely mystery. Many of us were left heartbroken at the loss.

My own work, concentrating on the fortunes of larger marine fossils that I had been collecting for ten years from the outrageously beautiful sea-cliff exposures in southern France and the Basque region of Spain, had reached an ending point at that time as well. Each year beginning in 1982, I had prospected among these most breathtaking coastal exposures of sedimentary rocks and combed the seaside strata for fossils of the long-ago Cretaceous Period. Slowly a list of species for these cliffs was erected, the cliffs and their sedimentary strata measured and cataloged, various sections correlated so that an intricate web of timelines stretched across a hundred miles of the Basque coastline. Each year I knew a bit more, and the picture of the extinction's starkness and speed became clearer. Scientific papers followed, and then a large monograph, a labor of love, publicly invisible but the stuff of progress in science, carried me as far east as Pamplona. The days were long but, in a not-so-strange way, anticlimactic. There was no more great mystery to be ferreted out of these rocks. The Cretaceous/Tertiary boundary sites in Pamplona were twins of those I had been working on since the early 1980s; the fossils told the same tale: a sudden and vicious extinction wiping out most of the marine species and all of the marine ecosystems. Like the sites I had studied near the Basque towns of Hendaye, Bidart, Sopelana, and Zumaya, the story never varied; there was little that could be added. As at the end of most major battles, now there was only mop-up work—small projects, things that were interesting but not world changing—and that was the problem, for the prior decade of work had been so exhilarating and addictive. By 1991 only a handful of doubters was left; the vast majority of those who had studied these rocky piles of strata of the latest Cretaceous age had no doubt that a great asteroid had hit the earth and that the end result of that collision had been death—death to the dinosaurs and plants on land, death to the ammonites and many other invertebrates in the sea, death in a hurry. The discoveries, the conferences, the pollination, the battles in public and print—none of us involved in the K/T-extinction controversy will ever forget it, and none of us will ever stop missing those heady days.

For me, there was one last day of work. My gear was packed, and I took a flight to London in order to wait for a visa from the South African embassy. I had to swear to them repeatedly that I was not a journalist. These were the last days of apartheid, and the great struggle that eventually toppled the long-oppressive white regime was nearing its culmination. The South Africans had no welcome mat out for journalists.

Thus, on a cool, late-fall night at Heathrow Airport in London, I checked my gear, boarded the South African Airlines jet, and headed to the other end of the world.

We cramped passengers flew all night, and early morning showed a dry countryside below. I was tired but ecstatic and excited: I had never been to Africa before. One door had closed, and another strange new door was opening. Customs was long and brutal, the agents all white and suspicious, searching, cold-eyed. The police presence was military and armed that way.

My South African colleague duly picked me up at the airport, took me to his house for a cup of tea, and then dropped me off at my hotel. It was, unhappily, a Holiday Inn. I could not stay inside this too-familiar vestige of the American life I had left, so I walked into the center of town and, in so doing, walked into Africa.

Tropical plants of deepest green, filled with strange squawking birds, lined the streets. I made my way into a central square where a huge daily bazaar was in full swing. There was a rhythm and a tempo to the life on the street that were distinctly new, and I was soon overwhelmed by the huge African marketplace economy bustling amid the grayer, drearier business buildings of downtown Cape Town. Music boomed, people bartered and thronged, there was bustle and motion everywhere—a visceral joy in living. And looming over all was Table Mountain, impossibly flat on top, a giant pile of strata towering above the city. It was mesmerizing. Cape Town, the self-proclaimed Mother City.

Africa.

From the St. George's Mall marketplace, I made my way

through busy streets into the Company Gardens, a more tranquil stretch of greenery, caged birds, fountains, and reflecting pools wedged between municipal museums. This huge garden gave way to my new scientific home, the South African Museum, which rose in splendor from the foliage, its turrets pointing squarely at Table Mountain. It was a Sunday afternoon, and Cape Town was crowded with kids and adults strolling through the gardens or visiting the many museums. A sharp, clear day, as so many in Cape Town are, the sky deepest blue and the wind blowing hard, sweeping away the smoke of the leechlike squatters' camps grasping the edges of the city. Table Mountain hung suspended over the museum, looking perhaps like Jupiter does from its nearby moons; even when it's not in direct sight, one knows that it looms there.

The museum was nineteenth-century ornate, a beautiful edifice and gift to the country. Sadly it was also nineteenth century in the ways of its exhibits. Long display cases filled with things old and older, great halls cluttered with aboriginal artifacts, and dioramas exhibiting Bushmen in their so-called native habitat in a fashion similar to those exhibiting the local animals in their native habitat. Nowhere, however, was there a diorama showing white people in *their* native habitat—or maybe that was the joke, as we white people walked through the museum staring at these representations of the "natives." Finally passing out of this sad tribute to racism, I entered the wings of paleontology, and here my pulse quickened.

Like all such exhibits, this one began with a treatise about time—geological time, that is. No other field of science has found it necessary to codify the timescale applicable to, and usually known only to, geologists. There is no formalized biological timescale or chemical timescale, although, of course, all processes described by these two great fields have temporal components. All other fields of study simply use the intervals of time known to us all: seconds, minutes, hours, days, and so on. Geologists, on the other hand, talk about periods and epochs, eras and zones, stages and series, the arcane subdivisions of what is known as the geological timescale. All are defined by death. The bigger the division, the greater the body count. For geologists, death becomes the ticking of the geological clock.

The divisions of time used in geology come from a study of the fossil record. Major time units are recognized and defined by mass-extinction events, sudden global catastrophes causing major biotic turnovers and extinctions. Two of these were especially dramatic. At the top of strata named the 250-million-year-old Permian System—and at the top of a much younger, 65-million-year-old Cretaceous System—the vast majority of animal and plant fossils were replaced by radically different assemblages of fossils. Nowhere else in time were such abrupt and all-encompassing changes in the faunas and floras to be found. These two wholesale turnovers in the makeup of animal life on earth were of such magnitude that they were used to subdivide the geological timescale into three large-scale blocks of time: the Paleozoic Era, or "time of old life" (extending from the first appearance of skeletonized life 530 million years ago until it was ended by the gigantic extinction of 250 million years ago); the Mesozoic Era, or "time of middle life" (beginning immediately following the great Paleozoic extinction and ending 65 million years ago); and the Cenozoic Era, or "time of new life" (extending from the last great mass extinction to the present day).

The two greatest of the mass extinctions also received attention in this musty exhibit. I was particularly interested in the explanation for the first of these two, the Permian extinction. I was not surprised to see the same cant that I had learned as an undergraduate: *By the end of the Paleozoic Era, some 250 million years ago, there was but a single "supercontinent," composed of a united North America, Europe, Asia, and Africa. Two effects of this gigantic, tectonic embrace supposedly produced the extinction. First: The earth's climate changed. Because of its immense size, huge areas of this supercontinent could no longer be cooled or warmed by steadying maritime influence, and the interiors of this gigantic continent thus grew hotter in summer and colder in winter. Second, when the contents coalesced, the level of the oceans fell dramatically, causing the wide interior seas found on virtually every continent at that time to disappear. It was within these shallow seas that most Paleozoic marine life had lived.*

The two processes were linked. As the climate grew more arid, the shallow seas evaporated all the faster, and with their loss the climate worsened, for these great inland bodies of water must have had an ameliorating influence on the climate. The earth's species gradually succumbed to the killing climate, slowly falling away like browning leaves, a few to be immortalized in rock, the rest to pass from all memory. By the end of 10 million years, only a tiny percentage of Paleozoic species was left, land and sea creatures existing in the few temperate refuges where great equatorial heat balanced frigid polar cold. A 10-million-year extinction!

I then moved on to the explanation for the 65-million-year-old extinction of the dinosaurs. Here, too, there was an explanation based on slow climate change and a story purporting that, like the animals killed off in the older of these two mass extinctions, the dinosaurs and ammonites also had died out gradually over millions of years. But at this time in 1991, we *knew* that this latter explanation was wrong: The dinosaurs might have died out in as little as a few weeks following the asteroid strike, not millions of years, perhaps not even tens of years. And if one of these stories was demonstrably wrong, why not both?

I moved on to the fossils themselves, those found in the nearby Karoo Desert in rocks from the end of the Paleozoic Era: the victims of the first of the two great mass extinctions. In poor light, amid dusty and yellowing backgrounds, I again found dioramas, those favorite devices of the old days of museology. But within these were creatures completely unfamiliar to me. Nothing in popular culture—or even in my long paleontological training— had ever led me to such images: giant hulking animals with ridiculous-looking mouths and the most ungainly carriage, huge front legs and short hind legs, behemoths eight feet tall and black in color. Monstrosities.

I looked at the fading script below: MOSCHOPS. A MAMMAL-LIKE REPTILE FROM THE ANCIENT KAROO. Near it were other monsters, some carnivorous, others herbivorous, all arranged chronologically so that the visitor could see how they changed through time. It was clear that these early land-living reptiles had indeed rapidly improved from truly clumsy-looking and obviously slow-moving behemoths to more agile and clearly more active forms. As I strolled through the cases, I was a bit chagrined to realize that I knew so few of these ancient creatures, whose names were largely foreign to me. Most of the figures inspired ridicule, a combination of their design and the way in which the models had been sculpted, being rather cheesily built and painted. In one diorama a particular monstrosity struck me. Clearly a carnivore, it was an image of a demon. On the bottom was a placard proclaiming this to be a gorgonopsian. Gorgon for short. The name conjured vague images from my painfully short and shallow training in the classics. *Some sort of Greek monster, a Gorgon. Three sisters, all nasty, Medusa the best known. Look at her and you turn to stone.* The name sounded sinister enough, and the creature in this musty diorama did indeed evoke a shiver: I had no wish to ever confront one of these things alive. Then a different image arose in my mind: *Chimera*, another monster from the Greeks, with the head of a lion and tail of a serpent. That seemed as least as fitting, and a better description of how this prehistoric Gorgon may have looked when alive. A novel monster, not the familiar lion or tiger of our world, nor the equally familiar and oversold carnosaur of *Jurassic Park* but a new nightmare or, more correctly, a new old monster. *Gorgon*.

▪ ▪ ▪

Thus was passed my first afternoon in Africa. Since I had come from Europe, far to the north but neither west nor east, I was in the same time zone and felt no ill effects from jet lag. I dined, reveled in the new smells and tastes, and exulted in the clear air Cape Town is blessed with, an air cleansed by the nearby sea. It is a place that inspires a distinct delight in being alive.

The next morning I was given a small office in the museum, met with my scientific host again, and was ready to begin work. I was there to study my main research interest, fossils called ammonites, which were shelled mollusks looking much like the modern-day chambered nautilus—we think. No ammonites survived the K/T extinction, and we are left with only their fossilized remains. I was interested in comparing the ammonite fossils found in the Southern Hemisphere with those I had collected in my native North, and in this I was not disappointed. The museum's collection of ammonites was indeed magnificent. After some days studying them, however, I began itching to see them in context—to see the actual rocks from which they had come, to journey to the field and see firsthand the outcrops of Africa yielding these specimens. Of course, I also wanted to find some myself, for I am a collector deep in my heart, like all true paleontologists. So I approached my host, Herbie Klinger, and asked directions to his field area, fully expecting him to offer to take me there. It turned out that the outcrops were many hundreds of miles up the coast from Cape Town, far into the tropics of Zululand and surrounded by jungle. In a most diplomatic way (and for a variety of reasons that were unknown to me at the time) Herbie explained that he thought it unwise for me to go there.

Here was a conundrum. If I wasn't able to work in the field, my options were limited. I could become a vacationer or try something else. Luckily, "something else" offered itself up relatively soon: one of the museum staff, Roger Smith by name, offered to take me to see Permian rocks in the Karoo. I knew virtually nothing about the Karoo and nothing about the Permian extinction other than its magnitude—that it was the greatest of all time. The party line in my field had been that the Permian extinction was a long-drawn-out affair, nothing at all like the very rapid Cretaceous extinction, and thus not very interesting. Nevertheless, the chance to see these famous rocks and look at a far older extinction than the one I was so actively working on was a chance not to be missed. I accepted with alacrity. And thus a change in my research direction was begun and a long partnership entered, neither of which I recognized till after.

■ ■ ■

Roger Smith is a presence. He is a head shorter than me, stocky, bearded, and weather worn. Yet he radiates an animal-like vitality, a coiled presence, affability covering careful menace. From the many photos scattered about his office, I realized that he was a distance runner of some sort. He seemed friendly enough, but he stared at me with very direct eyes, and I had the distinct impression that he was taking my measure. A very cool customer, Roger Smith. And strong—physically, mentally, emotionally. That was immediately clear. A strong man. Used to leadership. And used to living behind walls like all Brits, the heritage of privacy and stoicism.

On the appointed date, we set out, and my first surprise—shock, really—was how far away from Cape Town we had to travel to get to the Karoo. It took eight hours of fast driving to arrive at the area Roger suggested we visit, and this area was only partway into the vast expanse of dry land and semiarid desert known as the Karoo.

The drive was a journey through fantastically changing landscapes. Cape Town is located amid a rugged set of coastal mountains, and these continued inland for the first hour, a verdant green landscape of piled hills and upthrust low mountains deeply eroded. This was wine country, and for miles in either direction there were green valleys carpeted with lush vineyards. The mountains rose ever higher, and eventually we passed through a lengthy tunnel underneath the tallest peaks. When we emerged into bright sun after the long gloom, it was immediately clear that the country had changed for the drier. As we rolled on, hour after hour, the dryness increased, until Roger proclaimed what we were now in the Karoo.

The Karoo countryside is very distinctive, looking like no place I'd ever seen before. Most distinctive was the fact that all the low hills we passed had flat tops, called "kopjes." This is an odd shape for hills. Most hills are gently rounded on top, but not in the Karoo. I asked Roger about this, and he explained that the Karoo landscape came about through two very different geological phenomena that are virtually unknown elsewhere, thus giving the land its unique appearance. First, sedimentary rocks were deposited. Then these sedimentary rocks were covered and invaded by hot lava, creating a pastiche of intermixed sedimentary and volcanic rock.

Most of the rock around us was flat-lying sedimentary rock,

deposited more than a quarter billion years ago. Within this rock were the fossils we had come to see. The second type of rock here is dolerite, an igneous rock that makes up the tops of higher ridges in many regions of the Karoo. Dolerite sits like a great flat sheet upon an older bed of baked rock. It is the lithic equivalent of a cancer, an obdurate invader shouldering its way into the native country rock through sheer force of heat and pressure, melting or metamorphosing the sedimentary rock that sat there first. The dolerites made their way into the other rocks here more than 180 million years ago, a time when great dinosaurs ruled the earth; when huge *Brontosaurus* and slavering, dagger-toothed *Allosaurus* roared, roamed, loped, and ruled as the top carnivore always does; and when mammals were still merely cunning nocturnal vermin living in the cracks and crannies and tenements of the Jurassic jungle, the inner-city dwellers of the forests too poorly armed and not fleet enough to make a life in the open. The mammals were the weeds of the earth when the dolerites sprang into being, vermin to be sure, but tasty, with their warm-blooded flesh covered only by hair. And while all of this living mayhem ran its course, a greater event shuddered under the feet of the Jurassic world of Africa, for it was during the Jurassic that Mother Africa, like a great cruise ship, cast off its mooring from the giant supercontinent called Gondwanaland. Gondwanaland, earlier disinct, then merged into even bigger Pangaea. The signature of this colossal and messy divorce is the dolerite lining the top of so many places in the Karoo, bits of Hades rushing up to fill the cracks as the continents spread apart.

My most vivid impression from that first day in the Karoo was that it was a land of rocks, with all else an afterthought. This is not to say that there was not vegetation; there was abundant grassland and low shrubs, and even a gnarled tree on occasion. But the plants were only spacers between the rocks, which came in all sizes, from great boulders to small cobbles. Every hour we would get out to stretch our cramped muscles and take long drinks of water against the heat, and I would try to walk across this landscape. But, as I was to learn, walking across the Karoo landscape is an exercise in avoiding rocks.

For the next several days, we looked at outcrops as geologists are wont to do. I was particularly interested in seeing the boundary between the Permian and Triassic periods—an interval defined by the most catastrophic of all mass extinctions, the Permian extinction event of 250 million years ago. I was surprised to learn that no one could really put a hand out to any given outcrop and say, "Here it is." At the K/T boundary sites, there was no ambiguity—the position of the boundary clay was obviously present and eminently visible. Here in South Africa, in rocks that had been deposited by rivers and swamps, there was no such boundary. So I asked about the fossil record here in the Karoo. Another shock. Roger told me that no one had ever collected fossils from measured sections here. In other words, although thousands of fossils had been collected from Karoo rocks, and certainly hundreds that would not only be highly relevant in trying to ascertain not only *where* the extinction level in these rocks lay but also potentially yield important information about the *why* of the extinction among vertebrates, no one had yet conducted research of sufficient precision to attack this problem.

We had booked into a small suite of cottages in the town of Graaff-Reinet, one of the bigger towns in the Karoo. It was charming—Old Dutch architecture everywhere—and over a fine dinner, we planned the next day's activity. Roger suggested that we climb one of the odd conical hills (something akin to a mesa) that make up the Karoo landscape. On this climb we might pass through the Permian/Triassic-extinction boundary.

The next day we made the climb—and never saw a single

fossil. Roger assured me that they were indeed there, but I didn't see a scrap of bone. This was my first indication of how hard the scheme then hatching in my brain—a study of the ranges of the mammal-like reptiles across the boundary—would actually be to realize. We did see fossils at other localities, and sometimes huge numbers of bones, enough to whet the appetite of any bone hunter, but around the boundary between the Permian- and Triassic-aged rocks—known as the Permian-extinction boundary—there was nothing.

After two days of looking at various localities, we headed back to Cape Town, deep in conversation throughout the daylong drive. I had seen puzzling but enticing clues, things that made me realize that there was a better-than-even chance that the Permian extinction—at least here, among the mammal-like reptiles—was anything *but* the long, slow, drawn-out event that was portrayed in all the geology books of the time. Could it not have been as fast as the K/T event, perhaps even resulting from the same cause—a giant asteroid hitting the earth some 250 million years ago? I knew that, soon after the Alvarez announcement in 1980, there had been a flurry of excitement in China, caused by the reported discovery of iridium in a Permian-extinction boundary layer there—sure evidence of an asteroid impact with the earth. The Chinese came forth with this discovery in triumph. But that triumph was short lived. Retesting of these same rocks by other labs could not confirm the Chinese discovery.

Now, some ten years after that first, false announcement by the Chinese, there was still no consensus about the cause of the Permian-extinction event. But no one had ever looked at the Karoo rocks in any detail, certainly not in the way that we Cretaceous workers had studied the K/T boundaries throughout the world. No one had ever sampled the boundary for its fossils and geochemistry in such a way as to refute the possibility of an asteroid impact at the end of the Permian Period. The decade of work by so many of us on the number of K/T boundaries around the world provided a plan, a map, a working model of how to proceed here. And so, on the long drive back, I tried to instill this same sense of excitement in Roger Smith, who at that time worked on how rivers deposited strata and on how dead animals became fossils, but not on a far larger question: *What caused the greatest mass extinction of all time?* Perhaps the clues lay in the Karoo. Perhaps a huge scientific prize was ours for the discovering. Roger, it turned out, was a very good listener.

I came back to my desk at the museum filled with excitement. And ignorance. To do anything with the Permian extinction, I would first have to go back to school or, more appropriately, to walk the paths of those who came before, those who first brought the Age of Protomammals to life, those who named them and reconstructed them. Before I could understand how the animals and plants of the Permian died, I would have to know how they lived. I had to turn to books and weeks of study to uncover first clues to the secrets of the dry desert that had now worked itself under my skin.

Chapter 2

Bones in the Karoo

NOVEMBER 1991

The "Cape Doctor," the strong southeasterly wind coming off the southern coast of Africa, was making one of his frequent house calls, blowing away the smoke and pollution of humanity from the air above Cape Town. It was the Doctor that set the famous "tablecloth" on Table Mountain, piling billowing wet clouds so high on the flat tableland that they eventually spilled over the edge of the mountain like a falling cloth, a slow-motion white waterfall of cloud that flowed like dry-ice vapor. Torrents of it came down into Cape Town. The mist passed first through the trendy restaurants high on Kloof Nek, then past the colonial-era Mount Nelson Hotel and into the East India Company Gardens. It flowed by the legions of homeless asleep under the eaves of an art museum and the South African Museum, wisping by the statues of the great white robber Cecil Rhodes and the brutal soldier Jan Smuts, along St. George's Mall into the heart of the financial district, by the old Malay Quarters of District Six, flowing over the squatter camps and million-rand mansions alike, passing the whores lined up along Main Road in Sea Point as well as the models and their consorts strolling nearby Beach Road, finally slinking in tired, dissipated fashion into the sea along the grimy waterfront. This scene was a near daily occurrence for me as I struggled to come up to speed on the Permian extinction. I knew almost nothing of South Africa, and approximately the same about the Permian extinction. That changed. Both had now entered my life.

I lived in a hostel a block from the South African Museum and walked to work each morning. I was dating a South African woman and thus acquired Boer company, sweetness and sensibility, a car, and a guide to Cape Town and its customs otherwise impossible to gain. And I became a fixture in the library of the South African Museum, prowling stacks and records alike in search of an education. Asking a question entails knowing at least half the answer. I knew nothing and could not yet even pose a coherent question.

I had arrived at a historical turning point for this troubled country. A year earlier the ANC, or African National Congress, had been "unbanned" as a political party after decades of strife and violence. President F. W. de Klerk also announced in his bombshell speech of February 2, 1990, that political prisoners would be freed. The most famous of all was, of course, Nelson Mandela. On February 11, Mandela, who at that time had been incarcerated for twenty-six years under the vilest conditions, walked to freedom through the gates of Verster Prison near Cape Town. Mandela was driven to the center of the city, where a large crowd had gathered. There seemed hope that indeed a new era had arrived. But by the time of my arrival, more than a year later, the march toward an election for all the peoples of South Africa was still bogged down. De Klerk and the Afrikaans Party that had ruled with such fascism, terror, and murder for so long wanted power sharing with built-in protection for whites. The ANC advocated a simple winner-takes-all electoral system. Added complication came from the Zulu National Party (Inkatha), which vied with the ANC for dominance and, like both other parties, was capable of the most unspeakable violence in pursuit of its goals.

On a more local scale, I arrived in Cape Town to find a city beset by what came to be known as the "taxi wars." The main means of transportation in the city are minivans, locally owned and run by Cape Town blacks. Competition among the various taxi groups comes about by violence; gunfire and even hand grenades are used to "outcompete" other companies.

There were other surprises. Because of apartheid, South Africa had been ostracized for decades by the West, and tourists of any nationality were rare. The country seemed to have been cut off from the rest of the world intellectually as well as financially. The oddest quirks were apparent. The heavy-handed government had censored Western films and music, so the local radio stations played an endless selection of Carpenters music. I felt that I would go mad as anorexic Karen Carpenter crooned bad sugar from every mall and elevator. Even the drug habits of the country were bizarre. In a place where police control at borders was overwhelming and efficient, an endless supply of the sleeping pill Mandrax somehow flooded in from its manufacturing sites in Europe. The pills were crushed and mixed with the local pot, called ganja, and then smoked. The combination produces a zombielike state, is vituperatively addictive, and was available for next to nothing in any bar or on any street corner at prices that even the poorest could afford. At night, along every street, were hundreds of blacks, nodding out on Mandrax and ganja. If ever a sinister government wanted to control a population into somnolence, this would be the way. I ran this paranoid conspiracy theory by some of my new South African friends, and they looked at me as if I were mad (which may be the case). Since I am a product of the sixties in America, and thus conversant with the endless reports from that time that the CIA was funneling cheap, strong heroin into the United States for just such sinister crowd control, I still cannot believe that the South African government was not somehow behind this. One thing was certain: The streets were absolutely crime free, with heavily armed cops and doped-out street people allowing safe passage anywhere in the city. Other than the taxi drivers blowing up each other (and their customers), the Inkatha Freedom Party killing off numerous ANC functionaries and politicos, the white secret police taking out black politicians of any party, and Winnie Mandela reputedly murdering children, there was no crime.

Against this backdrop I went to work each day to educate myself about the Permian extinction to the point where I could actually go out into the field again and study it. Much of what I had to learn was history—scientific history, to be exact. I had to come to grips with time, for the entire debate over mass extinctions is one dealing with time. Was the extinction slow, lasting millions of years, or fast? And how to decide? First books, then the rocks themselves.

▪ ▪ ▪

The venerable library in the South African Museum was filled with information about our world as it was some 260 million years ago, during the Permian Period. It quickly became clear to me that the Permian world was far, far different from our own in so many ways: geographically, chemically, climatically, and biologically. The most startling difference may have been in the positions of the continents: All the major landmasses had united into the single supercontinent called Pangaea, which stretched from pole to pole. Instead of the familiar seven seas, there were two oceans, named Panthalassa and Tethys, restricted from each other by the presence of several small blocks of land, such as China.

There was much information about the extinction that ended the Permian Period. I wanted to start with the most recent summary on the extinction that was available. I found such a reference in my library in 1991, from a book then only a year old. In it, paleontologist Curt Teichert, the most revered and knowledgeable student of the great Permian event at that time, noted that many Paleozoic life forms disappeared toward the end of the Permian Period much as the musicians did during the last movement of Joseph Haydn's *Farewell* Symphony, one musician after the other taking his instrument and leaving the stage, until, at the end, none is left. A nice description of an event lasting millions of years.

Teichert was a patriarch of my field at that time, nearing eighty years of age. He was no one's fool, at the top of the paleontological pyramid, someone who had seen more Permian-extinction boundaries than anyone else living. He believed that no mass extinction took place at the end of the Permian and that the long biotic changes marking the transition from the Permian to the Triassic took place over a time interval from 6 to 10 million years in length. Here again was a reiteration of a long, slow event.

At first I found this article discouraging: Who was I to disagree with the great Curt Teichert? But further reflection raised doubts. Teichert may have seen many boundaries, but nowhere did it say that he had ever visited the Karoo. Teichert's view, that the mass extinction at the end of the Permian lasted 6 to 10 million years, flew in the face of what I had seen on my first Karoo field trip. Roger Smith and I had started our examination at the base of the last "zone" of the Permian Period in the Karoo, a unit of rock (and hence time) defined by its enclosed assemblage of fossilized mammal-like reptiles. According to Roger, at least thirty differ-ent fossil species, perhaps more, were known from this pile of strata. The succeeding assemblage of fossils, universally accepted to be earliest Triassic in age, had at most a tenth as many different fossils. Therefore, the mass extinction in the Karoo took place at the transition from this highest and most diverse Permian zone and the overlying basal Triassic-aged zone. How long did that take? If Teichert was correct, that time interval *was* between 6 and 10 million years. But Roger and I had driven from the base of the last Permian rocks to the base of the first Triassic rocks. This last zone of the Permian was about three-hundred meters thick, and somewhere in this unit was the mass extinction. The mass extinction thus took place in however long it took to deposit three-hundred meters of sedimentary rock—*at most*. But that was assuming that the mass extinction took place over the time it took for the entire zone to be deposited. What if the mass extinc-tion took place at the *top* of the zone? It may have been very rapid indeed.

Geologists have a pretty good idea about how long it takes for rivers to accumulate a given thickness of sedimentary rock. As-suming that ancient rivers accumulated sediments in the way they do now (a pretty safe assumption), one can arrive at a rough estimate of between 1 and 3 million years for the duration of the last Permian zone in the Karoo—and hence the maximum time it took for the Permian extinction to occur on land. Perhaps there was a missing interval in the Karoo rocks—a long period of time when no rocks were accumulating? But Roger Smith had searched for evidence of such a missing interval and could find none. Something was strange about the Permian extinction, at least as it was manifested in these African rocks. It surely oc-curred, and it was fast. My intuition was that a major discovery was here for the making.

By the 1950s the geological, and to a lesser extent the

paleontological, history of Africa at the end of the Permian was well known, for Africa had already served as a natural laboratory, showing the evolutionary origin of mammals and yielding important evidence demonstrating the reality of continental drift.

It was known that, some 350 million years ago, the southern portions of Africa had been covered by a great glaciation. Huge sheets of ice cloaked the land, in some places a mile thick, and inhibited any sort of terrestrial life. But continents do drift, albeit slowly, and in that impossibly long-ago time, Africa drifted northward into more tropical latitudes. The ice began to melt away, leaving behind great piles of gravel that consolidated into the sheets of sedimentary rock now found in widespread deposits at the base of what is known as the "Karoo sedimentary assemblage." As the land emerged from under the ice, it was colonized first by plants and then by animal life. The plants were tough, hardy vegetation such as horsetails, mosses, ferns, and club mosses, which must have slowly spread across the land in the wake of the retreating glaciers. It would have been a difficult, precarious colonization in a frigid and dry world, for few landscapes are bleaker or more forbidding than land newly risen from beneath glaciation. Very little soil is present, while great piles of rock and sand litter the landscape, geomorphic refuse left by the melting ice. But, millennium by millennium, the sun spent more time in the sky each year as the continent drifted northward; plants grew and then prospered. Their roots helped weather the rock cover into soil, and the organic refuse they left behind as they died became food for new types of life. Soon the plants were not alone, for animal life followed. At first these would have been only a few scorpions and insects, then ever more complex life: amphibians, later reptiles.

The first remains of land-living vertebrates are found from the deltaic, or river, deposits low in the Karoo succession. These fossils have been dated as mid–Permian Period, making them younger than the familiar fin-backed reptiles of Texas Permian deposits, such as *Dimetrodon* and *Edaphosaurus*, but older than the first dinosaurs, which were not found on the earth until almost 50 million years later. In this 50-million-year period, the Karoo remained a stable basin, a place of heavy rainfall and perhaps yearly floods, lush with vegetation, with trees that lost their leaves each fall and regained them each spring. The land animals of the ancient Karoo Basin flourished and multiplied in both number and kind, becoming the richest assemblage of late-Paleozoic vertebrates known anywhere on the earth. The Karoo is filled with their bones; at any given time, at this moment, untold millions of skeletons from that long-ago, faraway garden are baking in the African summer sun or cracking under the harsh winter frost, eroding, disappearing to dust in the vastness of the Karoo.

Various experts have long thought that the earliest Karoo vertebrates came from ancient Russia, for fossils not dissimilar to the oldest Karoo land vertebrates have been recovered there, from strata slightly older than any fossil-bearing Karoo rocks. But that view is slowly changing; sedimentological conditions in the lowest Karoo strata were not ideal for fossil preservation, and it may be that the earliest Karoo reptiles are as old as or even older than the Russian species. The Karoo vertebrates had to have ultimately come from someplace else, since they could not have originated in South Africa. The glacial climate in the ancient Karoo Basin was too harsh, the ice too thick. Slowly, ponderously, land vertebrates migrated into the Karoo from warmer places. Whatever their origin, once into the Karoo, they exploded into a bewildering variety of shape, form, and size.

The first glimpse of this ancient African fossil record was achieved in 1827, when a fossil tooth and then the skeleton of a prehistoric beast were collected from Beaufort West, a town in the heart of the Karoo. This discovery was announced in the *South African Quarterly Journal* in 1831. It was, however, soon overshadowed by the prodigious efforts of one man, Andrew Geddes Bain, who first demonstrated to the scientific world the great wealth of fossils to be found in Karoo rocks.

Bain, a Scotsman who migrated to South Africa in 1816, took up residence in the Karoo and soon became a hunter and explorer of note. When his trading camp was pillaged by Ndebele tribesmen in 1834, Bain barely escaped with his life. He became an officer leading black troops in an ongoing frontier war, then transferred into the military engineering corps and became involved in road building. In this endeavor he began to stumble upon fossil bones. When he read Charles Lyell's great masterpiece, *Principles of Geology*, he realized that his adopted African homeland could yield important information to the discipline of geology. Bain soon spent virtually all of his time searching for fossils. He had to rent a room in which to keep his burgeoning collection of fossil skeletons, for he could find no institution in the country willing to take them. In frustration he sent his collections to England, where the two leading anatomists of the day, Richard Owen and Thomas Huxley, immediately seized them.

Owen, the man who coined the word "dinosaur," correctly ascertained that the fossils sent by Bain were new to science but erroneously concluded that these ancient Karoo bones represented a sterile family of reptiles that resulted in no living descendants. He formally described them as the first examples of a new order of reptiles. Only later was it realized that this new group of reptiles was the ancestor of mammals.

Bain went on to collect many more protomammals and even established a family dynasty of fossil finders, since his son followed in his footsteps, collecting skeletons for the newly constructed South African Museum in Cape Town. Meanwhile the collections sent to the British Museum by the elder Bain languished for many years, their true significance overlooked; Owen and Huxley believed that mammals had originated from amphibians, completely bypassing any evolutionary detour through the reptiles. But in 1870 the American dinosaur hunter and evolutionist Edward Drinker Cope was able to examine one of the Karoo fossil skulls and arrived at a conclusion very different from that of Owen and Huxley. Based on this new evidence, he suggested that that reptiles, not amphibians, were the immediate ancestors of the first mammals and that the Karoo fossils discovered by Bain came from the first branch of this evolutionary lineage. In one stroke the Karoo fossils went from a sterile sideline to the main event in the search for the rootstock of mammals. Following Cope's discovery a great fossil hunt began in the Karoo, conducted by a variety of interesting characters. But the dominant figure in the search for the first mammalian ancestor was an intellectual giant with very controversial ways: Robert Broom.

Broom was a medical doctor trained in Scotland, but his real passion was paleontology. He arrived in Cape Town in 1897 with the express purpose of tracing mammalian ancestry back to the reptiles, and he proposed to do this by studying successions of fossils. Broom first supported his fossil hunting by practicing medicine across the Karoo, but after several years he joined a university faculty in a small town near Cape Town. He was a prodigious collector and writer, and soon he began publishing a long stream of scientific papers describing his finds.

Broom's research laid the groundwork for the Karoo biostratigraphy still in use today, and it was a major contribution. Employing the principle of faunal succession first discovered by the English canal builder William Smith, Broom proposed a series of faunal zones that break up the Karoo strata into short time units.

Because of the rapidity with which the ancient Karoo protomammals evolved, Broom was able to do this by describing rock units containing key index species of short duration. This pioneering biostratigraphy was perhaps Broom's most lasting contribution. But his studies also confirmed the earlier work of Edward Cope, who had first suggested that Karoo protomammals were younger than the fin-backed reptiles of America, thus making them transitional species between finbacks and true mammals.

Broom established a succession of Permian and Triassic time units based on the Karoo fossils that serves as the basis of our modern subdivision. He erred in many things, but in any scientific career spanning a half century or more, who does not? His most grievous error was in his concept of a new species, for to him virtually each new fossil find represented a new species type. Much labor has been required to prune these various names out of the scientific literature, allowing us to arrive at a more accurate census of the species-level diversity among Permian and Triassic mammal-like reptiles and true reptiles.

Broom's legacy lives on in the literature and in museums in the form of the many specimens he collected and named. But it also has a human side, for he left behind students, adherents, and champions, none more important than the Karoo-born and -bred paleontologist James Kitching, who, more than any other man, brought to life the Age of Protomammals.

▪ ▪ ▪

Kitching has known great fame. In the 1960s he made what might be the most important fossil find of the twentieth century. On a cold, rock-strewn slope in Antarctica, he discovered a specimen of the mammal-like reptile named *Lystrosaurus*. This same creature is perhaps the most frequently found vertebrate fossil of the Karoo. But this particular fossil, the first common animal of the Triassic Period, had never been recovered in Antarctica, and its presence there constituted a powerful geological proof that, 250 million years ago, Antarctica and Africa lay joined with the other southern continents into the single supercontinent named Gondwanaland. This huge landmass had then split apart into separate continents and drifted across the earth's surface like great stately cruise ships, each carrying its animals—and fossils—with it. In the 1960s many earth scientists believed that continents could drift. Kitching's find of *Lystrosaurus* in Antarctica constituted one of the proofs of what is now regarded as fact but was then still hotly contested. The Antarctic discovery was a triumph of paleontology, but it was built upon the foundation of the less celebrated, slowly accumulated understanding of the fossil successions in South Africa—the result of Broom's many years of research.

The Karoo fossil succession was like a series of dynasties, with waves of evolutionary faunas following one upon another. These dynasties were demonstrated only through the painstaking removal of skeleton after skeleton from the thousands of feet of accumulated sediments in the Karoo. Each of these fossil assemblages became a highly formalized unit of time called a zone. By mapping out these zones, Broom and Kitching were the architects of a biostratigraphy still used today.

Kitching and his generation of paleontologists had revealed an evolutionary history of the invasion of land by vertebrate animals. The transition from amphibian to reptile was one of the great evolutionary jumps in the history of life. Becoming free of the need for water in order to reproduce enabled vertebrates to colonize the land once and for all. The evolution of the amniote egg allowed this. All this history was discovered by the finding of fossil bones, in places like the Karoo.

The exact fossil group that is involved in the transition from amphibian to reptile is still unknown. The oldest fossils that can be reliably placed within the reptiles, from localities in Nova Scotia, are of early Pennsylvania age, or around 300 million years old. Yet, even at this earliest point in time, there are already two lineages of reptiles. One of these would give rise to the reptiles that we know on earth today. The second would give rise to the mammal-like reptiles and hence are the line that ultimately gave rise to mammals.

The oldest of the mammal-like reptiles, known as pelycosaurs, differed from the other primitive reptiles in several aspects. These creatures were probably predators, eating amphibians or insects. None were very large early in their history, with a maximum size of about a foot, though over time their dimensions increased. They were four-legged and not too dissimilar in appearance from the amphibians from which they arose. But their ability to live on land without having to return to water completely differentiated them from the amphibians.

The adaptation allowing these early reptiles to be free of water was an enormous evolutionary gift in a way. Many forms rapidly evolved, new shapes and sizes and body types, as they adapted to the numerous habitats available to land-living animals. The mammal-like reptiles grew in size, culminating in early Permian times, at least in North America, in the great sail-finned reptiles, such as *Dimetrodon*. These in turn gave rise to other, more advanced mammal-like reptiles, including the dicynodonts and the Gorgons.

■ ■ ■

As a student I had read many papers by Kitching. To me he was a legendary paleontologist, and, with some trepidation, I sought him out. But he was gracious to a fault and apparently overjoyed to receive a call from an American, for at that time in South

Africa, Americans were little seen. I was soon invited to visit him and, better yet, travel with him into the Karoo.

I again made the long drive from Cape Town into the Karoo and met Kitching in the town called Graaff-Reinet. From there we traveled to even smaller hamlets. We began our fossil hunting in the farming town of New Bethesda, the birthplace of James Kitching some seventy years prior. It was a trip back in time, not only to see ancient rocks but, as I soon learned, a trip of nostalgia. A playwright who had come to find solitude now occupied the little house where Kitching grew up. Once-spacious gardens nourished by wells now run dry were mere parched wastes; the numerous fruit trees and verdure of the town of Kitching's childhood were long gone. New Bethesda, formerly an oasis in the brown Karoo, was a bare memory of the land that had nourished the bodies and minds of its early-twentieth-century Afrikaans people.

On this day Kitching had nearly completed the circle of his own life. But as he strode across the New Bethesda river bottoms, the stoop of his back straightened, his gait quickened, a slow grin spread, his skin grew rosier, and deeply blue eyes emerged glinting from the clouds that had enveloped them. It seemed to me that rejuvenation flowed from the rocks upward into this man. Barren hills and the hodgepodge of New Bethesda surrounding the two sides of the wide riverbed were the backdrop, but Kitching was clearly seeing something else as he showed me fossil after fossil entombed in the strata we walked over. He had seemingly mastered time travel, for he described to me the ancient world that this sediment and its fossil bones had come from, painting verbal pictures of the long-ago world of the ancient Karoo.

He knelt and carefully brushed dust from a two-foot-long fossil backbone clearly etched in stone, each vertebral section gleaming alabaster in the sun, and then pointed out another and then another. The fossils were not at all rare. All had names, at the time meaningless Latin to me. Little did I know that over the next decade these names would become as familiar to me as the names of my best friends. *Dicynodon,* a cow-size herbivore. *Diictodon,* smaller in size and in appearance looking like a sprawling dachshund. Larger herbivorous forms such as *Paraiosaurus,* which were related to turtles but without shells. Scattered among these herbivores were the skeletons of carnivores. Some were small lizard-like forms, adapted to eating insects. But larger types were found as well, some truly terrifying in aspect. The largest and most ferocious were the Gorgons, whose namesake's visage, if gazed upon, could turn anyone to stone. It is now the Gorgons that have been turned to stone, but the giant teeth and almost snarling curl of the jaws bespeak terror, carnage.

The rocks gave a cross section of an ancient ecosystem. Much had been preserved, but surely much more was missing: the insects, worms, spiders, scorpions, most of the plants; the giant fauna and flora without skeletons that can never be preserved in the fossil record. This was the world that Kitching was seeing on our hot African afternoon, a slow, torpid, animal-infested river crawling across a vegetated valley, filled with sounds, smells, and movement—a once real universe that exists still in the minds of some intrepid time travelers like Kitching.

By the end of the day, I had learned much and had discovered a curious thing. While there are still many details to be worked out, the Age of Mammal-like Reptiles—as viable as any Age of Dinosaurs or any Age of Mammals—now lay delineated in stark detail, thanks to the study and collection of the Bains, Brooms, and Kitchings. But if an age had been defined, what about the end of that age? The pioneering South African paleontologists were far more concerned with life than with death. To them the question of extinction was far less interesting than was the question of origins, simply because all extinctions were at that time known to be gradual things and therefore neither easily studied nor even worthy of such study. They were just not interesting enough to waste time on. When asked, Kitching showed little curiosity about the Permian extinction. It was my generation, which seemed so preoccupied with death, that came to study it, redefine it, and ultimately understand its importance. This generation of thinkers and doers arrived, one by one, to the ancient South African killing fields.

The first of us to arrive was Roger Smith.

Born in England in 1950 during that country's version of the Baby Boom, Roger Smith grew up in rainy Britannia. He was a smart child and then a fan of Jimi Hendrix. He eventually attended university, but in the troubled times, and he graduated in 1976 to find no job available in his native country in his chosen profession, geology. He searched for overseas work and found two leads—one working north of the Arctic Circle in Canada and one from the South African Geological Survey in Pretoria, which was seeking an assistant to map and collect Karoo fossils. He was newly married, and he and his wife, Sally, weighing the prospects of freezing or boiling, boarded a plane for South Africa. Little did he know that his new position would entail both.

The young British couple arrived in Johannesburg and received several emotional shocks in rapid succession. First, and most alarming, they could not speak the native language, which is Afrikaans. This, coupled with the fact that at that time many white South Africans (and many more blacks) could not speak English, made a new and alien place even more daunting. They then discovered the reality of South African pay scales. Roger's new salary was fixed at five hundred rands a month, approximately one hundred U.S. dollars—a sum insufficient to rent suitable housing in the city. Since they could not afford an apartment, Roger's city-bred wife found out that she would have to accompany him into the bush of South Africa where Roger's new job would be based. But these were the small problems, the ones they had control over—they could learn a new language and work hard enough to improve their money situation—challenges facing most new immigrants. What they could not influence in any meaningful way was the reality of apartheid, at a time when the outside world was still quite ignorant of the real social situation in South Africa.

Roger and Sally Smith had arrived in a country torn by strife, mired in hatred, and cut off from the rest of the world. They had arrived in a country where, *in the same month of their arrival*, black people were not allowed on Cape Town beaches, an Indian doctor was forbidden from operating on a white child, and a white barmaid was arrested for serving a black journalist at her place of work. They arrived just four weeks before large parts of Soweto, the black township adjoining Johannesburg, rioted following the massacre of hundreds of students—*children*—by white police. "My patience is at an end," Police Commissioner J. F. Visser announced at the time. Children were shot down for protesting the policy that required them to learn all school lessons in the hated white language of Afrikaans. The Soweto uprising lasted three days, spread across South Africa like a wildfire, and became the trigger that started the eventual end of apartheid—an end still many years in the future. Prime Minister B. J. Vorster told his police to restore order "at all costs," and thousands of young blacks were arrested across the country—arrests immediately followed by what police called an "epidemic of suicides" in prison. How many murdered young men and teenagers ended up in unmarked graves next to South Africa's extensive gulags? Those who know now plead for amnesty.

In response to this genocide, large numbers of blacks fled South Africa for neighboring Botswana and Swaziland, where they were recruited into guerrilla forces and smuggled back into South Africa under the leadership of the African National Congress, the party of imprisoned Nelson Mandela. Roger Smith and his wife were now living amid this countrywide conflagration. This was no longer a short paragraph deeply buried in the *London Times*.

Roger Smith worked for the apartheid government as a field geologist at a time when the world was just awakening to the true situation in South Africa. He had assumed that he would be stationed in Pretoria, with occasional field trips, but he was in for another surprise. Ten days after arriving in South Africa, he was sent into the Karoo—indefinitely—to complete his job, which was collecting fossils so as to map out uranium deposits. At this time, and also unknown to the wider world, South Africa was building an atomic bomb and needed fission-grade uranium, just in case one of its black neighbors—or its own black population—became too militaristic. Unknowingly, Smith was now working for this effort.

He was given a large Ford pickup truck with a camper shell over the truck bed—the new home for him and Sally. It was a whirlwind. In a ten-day span, they moved from their comfortable English city life to living in the back of a truck, in an unknown desert, in a country undergoing revolution.

Their first trip into the Karoo lasted seven months—seven consecutive months among the tiny towns of apartheid-era South Africa, at a time when skin color was the only mark of status and could make one a target. The Geological Survey assigned Smith a crew of six black field assistants: four Zulus, two from other tribes. These assistants were experienced at finding fossils, and it was their job to teach Roger Smith his new trade as well as to help conduct fieldwork. They and their new supervisor never talked politics, but all were aware of the events swirling through South Africa while they spent each day looking for fossils and each night camping together in some new wasteland locale, a group of wandering vagabond geologists and a captive wife.

On his very first day in the Karoo, Roger Smith was dropped off in the towering canyons of sedimentary rock and told to look for fossils. While five skulls were found that day by his crew, Smith himself found nothing. Not on the next day either. Or the next. Each day his crew would find valuable fossils, and each day Roger would look and find nothing. His university preparation and his discovery of numerous invertebrate fossils in Britain were of absolutely no help in this new endeavor of finding Permian-aged vertebrate fossils in the Karoo. But his past training did bring him one solace that would have enormous repercussions for Karoo paleontology. To busy himself on the long, fruitless days, Roger Smith began localizing the fossils found by his crew with exacting precision on aerial photographs and topographic maps. He worked out where each fossil came from in the succession of strata. This had never before been done for Karoo fossils—not by Bain, not by Broom, not even by James Kitching—and twenty years later it would ultimately allow a high-precision recovery of fossils at the Permian-extinction boundary to take place.

Roger found his first Karoo fossil on his sixth day, a beautiful large skull. Presumably he rested on the seventh, or at least reacquainted himself with his wife. In his second week, the finding became easier, as he learned the visual cues that led to discovery. Near the end of this week, he made an extraordinary find. While walking along an outcrop, he noticed a fragment of bone sticking out of the hard sedimentary rock. Upon closer inspection he saw teeth—not the peculiar tusks of the plant-eating dicynodonts, but teeth that looked far crueler. He excavated more deeply and found a mouth full of teeth obviously evolved for ripping flesh, not plants. When the skull was removed from the rock, Roger discovered that he had found the remains of the most prized of all Karoo fossils—a Gorgon. It was an electrifying realization—and a clear message that there were treasures to be found and that he had the skill, temperament, and desire to find them. Things changed. Work became pleasure.

Each morning at eight he would head out, save for those days when the cold froze all the water and the gas in the stove. On such mornings two more hours would be required to allow the sun to thaw enough water for the start of the day. His wife remained behind in the camper shell, in blistering heat or frigid cold, to get through the day as best she could, with no human company. At the end of his ten-hour day, he and the assistants would return to eat, then sleep off their fatigue. After three weeks Sally Smith began crying hysterically, for no apparent reason in the men's eyes. For every reason. How much solitary confinement can any human endure? But she stayed.

Roger realized that his wife was cratering. He went to the nearest farmer in the region they were then working in, whose own wife, it turned out, spoke no English. The two women began to spend time together anyway, both isolated in the realm of the Karoo. They could not talk to one another, but they shared an equal misery of abandonment. Just having another human around in the desert vastness was a great comfort.

Roger, on the other hand, was now having the time of his life, experiencing that wonderful and all too rare gift of having discovered his true calling. A city boy born and bred, he found that he liked the spartan life he now lived. He looked forward to the long days, he embraced the challenge of finding fossils, and he applied himself to learning how to find them. He was good. Very good. His early experience of going six days without finding anything was a shock, and it never recurred. He was enormously competitive, and having his workers outdo him each day was not acceptable. He began to try to understand *why* the fossils where found were they were; he used his geological training to think about their occurrences. All of the fossils were being found in ancient river deposits, and he began to wonder if there were any parts of these deposits more fossiliferous than others. He was becoming proficient indeed, not just in finding fossils but in understanding why they occurred where they did. Soon he was far better than any of his crew, and, with long periods of time by himself to think about things, he began to view the Karoo in ways that no one had ever done before.

For their part the black assistants working for Roger got an education as well. Here was a white man quite unlike any they had ever encountered. He came with no baggage—no stake whatsoever in the long and violent history of white man and black man in South Africa. Roger Smith had no ancient relatives murdered in raids or lost in wars. He had no reason to distrust his assistants, and he believed in racial equality—a rare thing indeed in South Africa at that time. Here was a man who was not from a family that had stolen from their tribes or put down "rebellious" behavior. Roger Smith's workers also found that they could not outwork this man or outcollect him. A level of harmony was reached, but governmental regulations cast an unavoidable pall over the group.

Roger Smith soon found that he had an additional duty besides collecting and geological mapping. Because he was the only white member of the crew, he was in charge of keeping his assistants' pass books up to date. Of all the restrictive laws that made up apartheid, perhaps none was more hated than the Pass Laws. They were put in place early in the twentieth century to restrict the movement of nonwhites, and by the time Roger Smith arrived in South Africa, there were thirty-four separate statutes making up the Pass Laws. These became among the most powerful tools of white repression against nonwhites. All blacks and coloreds (the two official nonwhite racial categories) had to carry pass papers at all times and were not allowed to move out of a specifically designated region of South Africa without government permission. Any nonwhite caught outside his tightly controlled region was summarily imprisoned. For much of the century, imprisonment in a South African penal institution could be a death sentence, depending on the humor of the guards any particular week. These laws were perhaps at their most oppressive when Roger Smith arrived in South Africa, and, as a government employee, he was required to sign the pass papers of his crew each week and then, each month, travel to the nearest government center and have the pass books signed off on by a local official. There were no exceptions, and if he did not fulfill this obligation with regularity, his men would face arrest. Roger Smith came to hate these trips and what they stood for. But the task could not be circumvented in any fashion. The government was deadly serious about controlling the movement of nonwhites in the country, even between nearby towns. Black paleontologists crossing the Karoo in search of fossils were in no way exceptions.

For three years this went on. Months in the field, a week or two back in the city, and then back out again. Sally Smith became pregnant, and a son was born in the Karoo. Mother *and* baby now lived in the back of the pickup truck by day, with Roger in there as well by night. Washing diapers, cooking, eating—all was done in the freezing cold, the blazing heat of the Karoo. No electricity, no bathroom, no nearby doctor, no pharmacy. Nothing but desert. The Smiths shared the same degree of health and social care as the nation's seventeen million or so blacks—none.

The political backdrop stayed grim. During the Smiths' second year out, black activist Steven Biko was arrested in Port Elizabeth, tortured by the South African police while in his prison cell, and then, needing hospital treatment, was driven, naked, in the back of a Land Rover for fifteen hundred kilometers to Pretoria for medical attention, in spite of the fact that he was within minutes of a modern hospital in Port Elizabeth. He died on the stone floor of his cell in Pretoria, still awaiting medical treatment. The police claimed that his head was injured when he fell against his cell wall. News like this filtered through, yet Smith and his black crew continued the lonely work of geological mapping across the huge expanse of the Karoo Desert, spending every waking hour looking for fossils, collecting them, packaging them, and sending them off to the Geological Survey offices in Pretoria. Smith became tough and observant, inured to extremes of weather and isolation. He found that he loved heat, the hotter the better. Gradually he became the best in the country at this arcane skill, perhaps among the best in the world.

In his third year of Karoo work, Smith learned that Prime Minister John Vorster had been removed from office for stealing huge amounts of state funds and replaced by P. W. Botha, the former defense minister. The country braced itself. By reputation Botha was even more repressive than his predecessor.

Smith's work blossomed. He made fundamental discoveries and was confident that both his herculean efforts for the survey and the great personal price he and his family had paid to complete the work would not go unnoticed. It was therefore with calm satisfaction and expectation that he awaited his job review as his three-year term, agreed upon at his hiring, came to an end. Senior officers of the survey duly sent Roger his evaluation. Over the objections of his supervisor, Andrew Keyser, Smith was given a "second class" evaluation, which meant no promotion and no raise in salary. His work with Keyser became a monumental reappraisal of the previous work by Broom and then Kitching, and it is the basis for all modern-day work on Karoo stratigraphy. In spite of this, the survey found him lacking, not because of his professional work but probably because of his British nationality and because he did not speak Afrikaans. The key to promotion in South African civil service at the height of apartheid was to be a Boer.

Smith promptly quit. But by this time he was a bit of a legend in the world of South African geologists, and news of his resignation spread among them. A diamond-mining company interested in his knowledge and expertise immediately hired him. He was put in charge of finding diamonds for a large region of South Africa. This involved a substantial increase in salary, and for the first time Smith and his wife acquired a house, provided by the mining company, next to its mine. For four years Smith conducted geological studies for the mining company and lived in a small mining town. A daughter joined the son already born to the family.

But doing research for a mining company did not involve regular travel or daily work in the field. There were no fossils to be found in diamond exploration. Roger Smith began to miss his Karoo days and nights, and he looked for a way out of the rigid life dictated by the diamond company. This decision was eventually made for him. International pressure against apartheid was mounting in the first few years of the 1980s, and the company, an international rather than locally owned concern, simply divested, pulled out of the country, and laid off its entire staff, including Roger Smith. Smith climbed into his car, drove across the length of South Africa to Cape Town, and applied for a Ph.D. program at the University of Cape Town. He was turned down. Chagrined, he thought about other options. He had a family to support. He phoned paleontologist Mike Cluver, newly appointed as director of the South African Museum, explained his situation, and to his surprise was immediately hired on as a museum scientist for the princely salary of one thousand rands a month—then about two-hundred dollars U.S. It would be impossible to support his family on that sum, but he found that the apartheid government would provide him a house, because he was white. He made the long drive back to his house in the northern corner of the country. The family and all its possessions were loaded up, and within short order his life at the South African Museum began. He was allowed to simultaneously complete his Ph.D. at University of Cape Town.

With this new job in hand, Roger renewed his work in the Karoo and began frequent trips out into the field once more. The big difference now was that he was no longer doing projects dictated by someone else—he was on his own. Smith's primary interest was not in the ages of the fossils that he collected but in how they came to be fossils at all. He entered the field of *taphonomy*, which studies the process of fossilization. He began by studying modern rivers and how animals become trapped and then incorporated into the sedimentary record, and soon he began to learn where in river systems bone would most commonly be found. He studied the deposits of rivers both modern and ancient and learned how whole river systems became fossilized and turned to rock. He could now walk any deposit in the Karoo, and, where you or I would see sedimentary beds, he would be on the various parts of a modern river, such as its floodplain, point bar, or cut bank; he would know if the river was of the meandering type, such as the modern-day Mississippi, or a braided river, a type that occurs near glaciers and in mountains. He could assess relative size and velocity, back when the rocks were actually part of a river system. He could tell what the source rocks had been, what kind of vegetation and soil the river had held in its banks. He learned to look at ancient strata and immediately know where the best place would be to search for fossils. And he could spot any fossil to be found, as if by magic.

At the South African Museum, Roger Smith had joined a staff with an extensive tradition of Karoo paleontology, going back in time in an unbroken string of curators to old Broom himself. A few years after Smith's hire, the museum would hire a second British paleontologist with an interest in the ancient Karoo world and it long-dead inhabitants, Dr. Gillian King, who would go on to become the world's authority on dicynodont protomammals. Gillian was another sixties survivor, about Roger's age, but she had arrived at this point via a different social and educational route than the one taken by class- and status-bedeviled Roger— and with vastly different training. Roger was a field man from Manchester who came to know the anatomy and osteology of the mammal-like reptiles as well as any expert did. Gillian was an anatomist from Oxford, untrained in things geological. As any Brit can tell you, it is a *long* way from Manchester to Oxford. The two should have been highly complementary, and could have formed a partnership of extraordinary talent and effectiveness. But for whatever reason, together they were like oil and water. By the end of the 1980s the Karoo paleontology group at the SA Museum was making headway in understanding many aspects of ancient Karoo life and how it came into the fossil record, but not as a team. Although Cluver and King joined forces to revise and sort out the taxonomic mess left behind by Robert Broom, the re-sult of his curious penchant for defining a multitude of new species, King and Smith never published together in any substan-tive fashion. Soon even their interests in Karoo paleontology began to diverge. Roger remained fixed on how animals become fossils, while Gillian King became increasingly interested in the first step of this process—how did things die off in the first place—especially at the Permian-extinction boundary. She began to publish papers about the Permian extinction in the Karoo. This work had been unknown to me in my vast ignorance of all things dealing with the Permian extinction, but I found them soon enough during my long days in the South African Museum library.

During most of my 1991 visit to South Africa, Gillian King was in Europe and was expected to return only in my last few days there. I very much wanted to meet her and ask some pointed questions about how she came to believe that the Permian extinction was a gradual rather than catastrophic event.

By the time of my arrival King had published three influential papers about the nature of the Permian-extinction event in the Karoo. And since the Karoo at that time was the only Permian-extinction section sufficiently well known from a nonmarine setting, where vertebrate fossils were abundant enough to also allow any sort of analysis, her papers took on an added importance and became proxies for how the extinction unfolded for vertebrates the world over. In this way Gillian King came to be the world's expert on the Permian extinction as manifested on land, and on its effect on Permian-age vertebrates.

At the time that King began this work, her colleagues in vertebrate paleontology elsewhere in the world had also weighed in that the mass extinction among the world's mammal-like reptiles and other vertebrates of the late Permian was more of a whimper than a bang. Paleontologist C. W. Pitrat, in 1973, analyzed the known record of vertebrate diversity across the Permian-extinction boundary and concluded that there was no mass extinction among vertebrates; the mass extinction affecting the rest of the world's organisms thus completely bypassed the vertebrates of the time. Pitrat believed that the Permian extinction involved only marine organisms.

Following Pitrat, paleontologist E. C. Olson reexamined this question in 1982, finding a slight reduction of vertebrate diversity crossing the Permian-extinction boundary, but nothing that would argue for a mass extinction. These two works by respected paleontologists went a long way toward convincing the paleontological world that little of import occurred on land at the Permian-extinction boundary. Both of these studies addressed the Karoo and suggested that what minor change there was could be attributed to the fact that the nature of sedimentary rocks in the oldest Triassic strata was different from those in the immediately underlying Permian strata. This convinced these workers either that the nature of preservation changed across the Permian-extinction boundary or that there was a missing interval at the boundary itself.

Against this backdrop of institutional acceptance of noncatastrophe, at least on land, King weighed in with a more sophisticated look at the Karoo vertebrate record than that attempted by either Pitrat or Olsen. She tabulated the number of Karoo vertebrate taxa for the Permian and Triassic and found that there was a steady decline in numbers throughout the late Permian, with no great extinction at the boundary itself. In one of these papers, she concluded that any hypothesis for diversity change that demanded an abnormal period of extinction at the end of the Permian was not backed up by data from the Karoo Basin—in other words, there was no extinction at all.

With King, the person closest at hand, attesting to a noncalamity at the Permian-extinction boundary in the Karoo, the rest of the paleontological world seemed content to agree that nothing much had happened (on land). This view was institutionalized by the world's overall expert on the Permian-extinction boundary, Smithsonian scientist Douglas Erwin, in his influential 1993 book *The Great Paleozoic Crisis: Life and Death in the Permian*. Erwin's discussion of fauna at the end of the

Permian concluded that there was no evidence for a catastrophic extinction on land.

As a scientist sitting in my armchair reading these accounts, I might have been perfectly satisfied that all of these eminent scholars had gotten it right. But I had just come back from seeing these sections. I had seen the imprecise record keeping that the old lions like Broom and Kitching had used for locality information; they at best localized their fossils as coming from a given farm, but never in terms of time. Thus one could never really know where a given fossil came from, either in space or time, for saying that fossils came from a ten-square-mile farm was akin to giving no geographic information at all. With such poor localization, how could scientists like Pitrat, Maxwell, and King, who did not even go into the Karoo to look for fossils or know about their actual locality of discovery, make *any* sort of sweeping statements, no matter how sophisticated their library analyses? And even more damning (in my eyes at least) was the experience I knew of firsthand, from my work on the K/T extinction. The vertebrate paleontologists studying the K/T extinction had opposed *any* theory of rapid extinction, but had never gone out and done the careful field analyses necessary to show whether dinosaurs disappeared gradually or suddenly at the K/T boundary. I knew from sad experience, based on my own work on ammonites at the K/T boundary sections, that sudden extinctions, at least in the rock record, almost always look gradual. I suspected that the same thing could be happening in the Karoo.

Roger and I came to a decision. Although it was now textbook knowledge that the Permian extinction among vertebrates was supposed to be a nonevent, we thought it a good gamble to retest that assertion with some high-precision field collecting across several boundary sections we knew of in the Karoo. We would try to find funding to attempt just such a detailed examination in the

Lootsberg Pass region. My thought at the time was to send a graduate student to live there for a month and scour the slopes across the boundary interval to arrive at a detailed picture of the fates of various species of vertebrates in the Karoo at the Permian-extinction boundary. Roger would provide vehicles, and he and I would come back as well to get the project rolling and help as field assistants at the start of the project. A small proposal was put together and hurriedly submitted to the United States' National Science Foundation in their "risky research" protocol.

As luck would have it, Gillian King came back just as we had finished the proposal. I was very much looking forward to meeting her and enlisting her aid in this project. Because she knew the taxonomy of the fossils we were about to go get, her help would be invaluable. It was a surprise, then, to find that she intended to do just the work that Roger and I were proposing. We were quite firmly told that our help was not required.

Dilemma. This was not my country, and I had learned through sad experience that it was never good to get involved in another institution's internal politics. This was something that Roger and Gillian would have to work out. I booked a flight out of Africa. Upon leaving, I supposed that I might never again visit this place.

Roger became inscrutable. He turned further inward with regard to Gillian and made his own plans, began keeping his own council. But seeds had been sown.

Chapter 3

Gradual or Sudden?

Gillian King never did follow up on her proposed research in the Karoo to sort out the Permian extinction among its fossil reptiles. She soon moved back to England, leaving the field open to Roger Smith. Roger began the first steps that led to a revolution in our understanding not just of the Permian extinction but of all extinctions.

In September 1992 a small group of influential South African paleontologists convened for an extraordinary field trip. The trip was an attempt to bring together the various Karoo fieldworkers, and to map out new research directions. Roger successfully lobbied to look more carefully at the Permian extinction and was joined by my new friend James Kitching, as well as by paleontologists Johanne Wellman, Bob Brain, and Bruce Rubidge, all from universities in Johannesburg. Kitching was in failing health, and this trip was equally an attempt to pass on the knowledge of a lifetime to a younger group as to start a concerted effort to understand the Permian extinction in the Karoo.

The group made a rapid tour of the Orange Free State, and early on, Kitching pointed out a hill where he had decades before collected undoubted Permian fossils at the bottom and Triassic fossils at the top. A Permian-extinction boundary had to lie in between. At this time it was Karoo lore that there was *no* place where an unbroken (thus complete) record of sedimentation across this boundary existed, even though Roger Smith already had expressed grave doubts about this well-known "fact." Kitching told the group that a missing interval lay in the middle of the hill, as it did everywhere at this time interval in the Karoo. Published papers by Kitching and many others stated that at about the time of the Permian extinction there was a break in sedimentation, meaning that large intervals of time were not represented by any rock and thus could not be used to test whether the extinction was fast or slow—since the critical evidence either was no longer there, or had never been there in the first place! According to this model, the boundary could not be a true record of events. Yet as Roger Smith stared at this hill, examining the nature of sediments with his keen and practiced eye, he realized that the way the sediment lay piled bespoke continuity, not breaks. He was an acknowledged expert on the sedimentology of river deposits, and all his decades of experience told him that this place was continuous—and thus an ideal place to study the Permian-extinction boundary.

Roger turned to Kitching and posed the following question: Where in the Karoo would be the best place to study the transition of fossils across this boundary? Kitching answered immediately: the Fairydale Farm, near the small town called Bethulie, in the Orange Free State on the banks of the Caledon River. And with that the entire field-trip crew, now fired up and on a mission, switched gears and headed toward Bethulie, changing the course of future events in South African geological research.

How Kitching ever found this site is itself a mystery. The

Karoo is a vast place, and the chance that someone would stumble on a site like Bethel Canyon seemed remote even for someone who spent as much time looking for fossils as Kitching did. It is a difficult spot to get into, a long hike from any road, and in the days that Kitching was doing his major work, there probably were no farm roads that allowed any access. But find it he did, and in doing so he found the most fossiliferous section of rocks around the Permian-extinction boundary that still, to this date, has ever been discovered in the Karoo.

The group, well equipped with four-wheel trucks, followed Kitching's directions, slowly making its way along a rough path to the edge of a giant canyon, finally stopping at a canyon's edge that faced a huge, panoramic view. Kitching was in no shape to make a climb down this canyon to the Caledon River far below, but he pointed out to the group the various places where he had found fossils and gave them a rough sense of where the Permian-extinction boundary might be located. Kitching's memory was extraordinary, and he could remember in detail each fossil that he had collected here, even those from decades past.

At this point Roger Smith turned to Kitching and asked if he could come back with a crew and make a comprehensive collection of fossils as well as an examination of strata in the canyon. Kitching, to his credit, had no objection. Paleontologists are a very territorial bunch, and this was prime Kitching territory. He had done much of his life's work here, but he was no longer physically capable of conducting this new type of precision fossil collection, and he had the good sense to know it, unlike many, who hopelessly cling to old field sites long after they have neither motive nor ability to continue working them. On this day Kitching passed a torch, to the ultimate benefit of South African paleontology and especially to those interested in mass extinctions.

Some months later Roger Smith did come back, and he brought

his field crew with him. Early on, Smith had learned the utility of having many eyes on the outcrop, and in the South African Museum, he had set out to accumulate a talented cadre of field collectors. By 1992 he had such a team. They approached the project and section at the bottom, beginning on the Caledon River, and then slowly worked up in elevation into ever higher, hence ever younger, strata. Even on their first day, they realized that they had struck paleontological gold. The banks of the river were composed of great expanses of sedimentary rock—an irony, having the banks of a modern river made from the stones of a 250-million-year-old one. On that first day, Roger immediately found the remains of several *Dicynodons,* the most common herbivore of the latest Permian, and then a *Theriognathus,* a dog-size carnivore. Near that fossil were the remains of *Diictodon,* a smaller herbivore.

This was very encouraging.

But collecting fossils is only one part of this work. More exacting—and time consuming—is measuring the stratigraphic section. Using a tape measure and a 1.5-meter-long staff graduated into 10-centimeter lines like a long ruler, Smith made a table of the strata in this canyon, starting at the base and moving up. He described each bed in exacting detail, listing its grain size, sediment type and color, and the nature of any sedimentary structures. It took ten days to complete this job, and all during this time Roger's crew was busy collecting and noting fossils. By the end of the ten-day trip they had collected more than fifty skeletons and skulls, a huge number for any single region. As advertised, this site was extraordinary.

On this trip, and on a subsequent one in 1993, it became clear that Roger's first appraisal of the section was correct—that there was no evidence of a break at the Permian-extinction boundary. Furthermore, Smith was able to demonstrate a relationship among fossils that had been suspected, but never proved: that

Lystrosaurus, the long-used paleontological marker for the base of the Triassic, first occurred in late Permian-aged rocks, not Triassic-aged ones.

The use of *Lystrosaurus* as a marker for the base of the Triassic System dated back to Broom, and it was still used as late as 1993, even though no one could find any evidence of mass extinction in the beds containing the first fossils identified as *Lystrosaurus.* Roger Smith, with years of Karoo experience, was skeptical. Some previous workers, most notably American paleontologist Nick Hotton, suggested as early as 1967 that there was an overlap between *Lystrosaurus* and the Permian *Dicynodon*-zone fauna, but it had never been conclusively demonstrated. Roger Smith now had that proof. The result from his fieldwork demonstrated that paleontologists had been searching the wrong beds to find the mass-extinction horizon. Smith began to look in younger beds.

If the proof of the overlap zone was surprising, Smith's next discovery was totally unexpected. As he climbed up the sedimentary section, he noticed a very peculiar change: The nature of the sedimentary bedding and structures showed a huge difference on either side of the Permian-extinction boundary, which was now defined as the bed containing the last occurrence of the mammal-like reptile *Dicynodon.* Sandstone bodies below the boundary showed the hallmarks of a single Mississippi-size, meandering river. But the beds above the boundary were radically different. Their structure suggested that no single river dominated the floodplain but that there had been many small ones and that these rivers were probably braided, the type of river and stream systems seen today next to glaciers or alluvial fans.

Smith ultimately took several more trips to this particular area and finally judged that he had enough material to publish a paper about the sedimentology of the region. The paper appeared in 1995 and remains a seminal contribution, perhaps the most important of Smith's long career. In it, Smith demonstrated the existence both of an overlap zone and of vastly different types of river systems above and below the Permian-extinction boundary at this section near Bethulie. To this point all was well and good. But in one figure of his study, Smith showed a radical new view of the extinction itself. The figure shows the stratigraphic ranges (their positions in the canyon's pile of strata and hence their ranges in time) of fourteen reptilian vertebrae genera found in the Karoo Permian rocks. In this diagram Smith showed three of these species surviving into the Triassic and nine of them going extinct right at the boundary. The data used in this figure are stated to come from his own collecting and from a forthcoming publication edited by Bruce Rubidge, then still in press.

This small figure represented a revolution in understanding of the Permian extinction in the Karoo. If correct, it overturned a century of belief that the extinction on land was gradual. However, the data presented in Smith's 1995 paper only partially support this chart. Since he used *Dicynodon* to pick his boundary, somewhere the data should list the last occurrence of the other fossil reptiles (*Procynosuchus, Youngina, Theriognathus, Pristerodon, Diictodon, Cyanosuarus, Prorubigei,* and *Rubigei*) as being recovered right at the Permian-extinction boundary bed, the same bed from which the highest *Dicynodon* was recovered. Perhaps these last occurrences are found in the paper edited by paleontologist Bruce Rubidge, a compilation published by the South African Geological Survey of the ranges of mammal-like reptiles throughout the Karoo? But this paper has no data at all, only unsubstantiated and highly generalized range charts. Years later Roger admitted to me that the figure in question was based partly on belief rather than on data. As it turned out, he was correct, but learning that would take many years of work and would drag me back into the fray.

Smith had now proposed the previously untouchable conclusion—that the extinction among vertebrates happened quickly rather than slowly—and he proposed a new explanation, one backed up by data—that because the extinction coincided with a radical change in floodplain morphology and regional climate, it may have been caused by this change. He concluded that the Permian extinction in the Karoo was caused by environmental alterations and the onset of severe drought conditions. This in turn caused the rivers in the area to change style. The permanent meandering rivers of the Permian became the ephemeral braided streams of the Triassic.

▪ ▪ ▪

Smith's 1995 paper was of enormous importance in a debate that at the time was rapidly heating up. In the same year that Roger Smith presented his revolutionary new view of the Permian extinction on land, other extremely significant papers were also published. Among them was the report in *Science Magazine* that geochronologist Paul Renne and his group had dated a Permian-extinction boundary in southern China using high-precision dating methods and showed that the age of the extinction was 250 million years, plus or minus 300,000 years, a nice round number indeed. They also noted that this date was synchronous with the largest eruption of flood basalt then known on earth, a pile of Permian- and Triassic-aged lava known as the Siberian traps. Flood basalts are volcanic eruptions of regional extent. They pump gigantic volumes of carbon dioxide into the atmosphere— which could lead to global warming as well as produce sulfate aerosols and massive acid rain. Here, then, was a second potential culprit in causing the extinction.

Also published in 1995 was a report by Yoram Eshet and two colleagues of the Geological Survey of Israel about a curious discovery made from Permian-extinction boundary sections in southern Israel. Petroleum-exploration companies drilling boreholes in southern Israel had recovered from their core samples strata that spanned the Permian-extinction boundary. None of this rock was available on the surface, unfortunately. Nevertheless, a detailed examination of pollen and spores recovered from the boreholes revealed something astonishing: A sudden and catastrophic event had occurred among land plants right at the Permian-extinction boundary. A sharp break in the floral assemblages was recovered, and this break was marked by a short-lived event characterized by widespread fungal remains. The sudden appearance of fungi corresponded with an almost complete disappearance of plant life; it was concluded by the authors that this marked a short-term but extraordinary catastrophe. Eshet and his group estimated that the narrow stratigraphic horizon with the fungal "spike" was deposited in less than 50,000 years—perhaps much less, for their estimate was a maximum duration. The catastrophe therefore was a relatively brief event by geological standards, far briefer than the 10-million-year estimate accepted as late as 1991.

Not only was there new work looking at the land animals, but by the mid-1990s there was important new information about how marine organisms fared at the Permian/Triassic boundary. Work by Douglas Erwin, Sam Bowring, and their Chinese colleagues showed that the tempo of marine extinction in China was appearing increasingly catastrophic and rapid at the Permian-extinction boundary sections there.

The century-old theory that slow climate change was the primary cause of the great Permian extinction was dead. A more rapid cause of the extinction had to be found. While many workers invoked the effects of a monumental outpouring of basalt in ancient Siberia, death by asteroid impact was still in the back of many people's minds. A better sense of the extinction's rapidity was needed. Work in China and Africa would concentrate on that question.

Chapter 4

Land and Sea

By the latter part of 1992, I was quite resigned never to see the Karoo rocks again or to have any part in the scientific problem that was unfolding with increasing excitement and challenge. I could only sit on the sidelines and watch as discovery after discovery came to light about this greatest of extinctions. And then, in one of these odd turns of chance, Africa became a possibility once again.

By late 1995 one of my former graduate students was about to finish his postdoctoral position but had no job waiting; he faced either a change of career or some serious starvation. I promised him I would see if I could come up with at least a one-year position of some sort, somewhere, and in so doing I set off a series of events that inadvertently took me back to Africa and the Karoo.

Ken MacLeod's dissertation research had been on the K/T extinction—he had worked with me during the late 1980s on the beautiful coastline sections in Spain and France. Subsequent to earning his Ph.D., he received a distinguished postdoctoral fellowship from the Smithsonian Institution. He spent three years there, applying for academic jobs all the while. Yet, as he was interviewing, the very tight job market seemed to have closed most doors. I pondered some project that might be palatable to a new funding program being offered by my university, a scheme wherein moneys collected from the discoveries of its professors were in a small way returned to investigators on a competitive basis. Searching for some topic, I thought once again about a project looking at the tempo and nature of the P/T extinction on land, based on research in the Karoo. Ken and I pondered: What could be done, especially in light of Roger Smith's new discoveries? And would Roger welcome me back? Ken suggested that rather than look solely at the pattern of extinction based on the ranges of fossils, a proxy be used: We could test for an isotopic "perturbation."

Isotopes are elements that vary slightly in mass. Carbon has three such isotopes, the very common carbon 12 (notated as C_{12}), the much rarer carbon 13 (notated as C_{13}), and carbon 14 (notated as C_{14}), which is used to date archaeological relics. Studying isotopic ratios from rocks and fossils has long been a tool for deciphering the severity and tempo of mass extinctions. One of the most persuasive lines of evidence are analyses of the ratios of carbon 13 to carbon 12.

Carbon is the major construction element that makes up all of life, including us. Most of our cells are composed of carbon, and most of that carbon is C_{12}. It turns out that because of its lower mass, C_{12} is easier to use by living organisms than is either C_{13} or C_{14}. Because of this, organic carbon is enriched in C_{12} relative to C_{13}. When biological productivity is high (meaning that a great deal of the carbon derived from the atmosphere or ocean is being taken up by plants and turned into living tissue), the ratio of C_{13}

to C_{12} is high. When productivity is low (usually meaning that little photosynthesis by plants is taking place), the ratio lowers. During times of mass extinction, life is killed off, especially plant life. When there is less life around taking up C_{12}, it becomes enriched compared to its two, more massive isotopic cousins.

But all this dancing with carbon happened a long time ago. How could we measure the ratios of isotopes that existed a quarter billion years in the past? It turns out that the ratios of these carbon isotopes can be preserved in both rocks and fossils. Since these carbon atoms are taken up by living tissue, and sometimes preserved in bone and shell, they can be sampled long after the living creature containing them has died. By sampling ancient sediments containing fossilized bones, leaves, wood, and shells, chemists can determine the periods in the past that were rich in life and those times when there was less, even very little, life. It is just this ratio of the carbon atoms that is examined at mass-extinction boundaries. The amount of depletion, and the amount of time necessary for one carbon isotope to deplete into another, gives two important clues—it indicates both the severity of the extinction and the rate at which it occurred. There is an additional benefit of doing these analyses. When a major isotopic perturbation occurs, it usually happens all over the world at the same time. Finding one of these excursions thus becomes a means of showing contemporaneity of rocks in far distant parts of the globe or in rocks deposited in very different types of environments.

One of the results from isotopic analysis at the Permian-extinction boundary was a first clue that this extinction may not have been as long lived as previously thought. In 1988 an obscure publication by Polish geologists described a major geochemical change in fossils found in Paleozoic rocks from the Alps. These scientists sampled the shells of brachiopods, shellfish common during the Paleozoic but largely extinguished by the Permian extinction. By analyzing the shell chemistry of these fossils, collected from a thin stratal succession, this group showed that, perhaps a million years before the end of the Paleozoic Era, a sudden spurt in productivity occurred, followed at about the Permian-extinction boundary by a decline so profound that it speaks of oceanographic change and a virtually unprecedented loss of oceanic productivity. In their view it could have taken place over a period as long as a million years or over less time—perhaps much less. In fact, such an isotopic shift has been seen on this scale at only one other time in earth history: at the Cretaceous-Tertiary boundary transition, the time of the K/T event.

The problem with this first study was that it was incomplete—no one knew if the strata that had been sampled formed a continuous sequence spanning the critical interval of time or if there were stratal beds missing at the boundary itself. A second study was undertaken on a core from the Austrian Alps and was published in 1991, headed by University of Oregon geochemist Bill Holser. This group was able to corroborate the first study, but in far more detail. Soon similar isotopic excursions were found in many sites, including very complete marine sections found in China.

But what of the land? The studies in the Alps and China described only oceanographic change, but the isotopic tool had not yet been tried on sedimentary sections that had been deposited in nonmarine settings; thus no one knew if a similar trend would be found. The first attempt to see if a similar perturbation could be found in a nonmarine section came from the Karoo. In the early 1990s a group that included Roger Smith looked at mammal-like reptile teeth from the Karoo but could find no trend equivalent to that seen in the oceans. Although they found a long-term decrease in C_{13} relative to C_{12} in their samples, there was no spike equivalent to that seen in the marine sections sampled by other geologists. The Karoo results suggested that the extinction was not a rapid event on land but was instead slow. Yet this study had some inherent flaws: The samples were not taken from measured stratigraphic sections, as the example from all other isotope studies had been, but were from isolated museum collections. Each sample thus had a different burial and heating history, and because of this the results could easily have been flawed. Ken and I thought that we might profitably redo this study by sampling from actual stratigraphic sections. We seized on this and proposed a new study. A grant proposal was submitted, and, to my gratification it was funded. I had enough money to pay Ken for a year and to send the two of us to Africa for a single visit.

As things turned out, Ken went by himself. In 1995, I had married, and by early 1996 my wife was pregnant, so I stayed home. Ken was none too pleased when, only a month before departure, he was confronted with the fact that he would have to go and do the fieldwork alone. I feared a bit for him but thought that my friends in the South African Museum would set him on a successful path, which is just what happened. There were certainly reservations and room for worry, however. How would Roger Smith react to this young scientist, coming at my behest and in my place, these many years later? Could he find the material necessary to conduct his analyses? Ken thus set out into unknown territory, unfamiliar with the place or the people, uncertain about his reception, unsure about the nature of the rocks, and not even knowing if the problem was tractable. Yet go he did, stopping first in Kenya to see the game parks and then heading south to Cape Town.

Ken spent a week in the South African Museum collections, learning the names of the creatures he would soon be seeing in the field and trying to get as organized as possible with maps, localities, and the other vagaries of the upcoming trip.

But he had hatched an ingenious plan. He needed organic material of some sort to seek an isotopic perturbation that might—just might—tell us whether the mass extinction on land took place at the same time as it did in the oceans. Yet the Karoo was remarkably devoid of rock with internal organic material. There were no coals or organic-rich rocks such as oil shale or even deposits of woody material that might yield a rock with sufficient organic content to allow its testing with a mass spectrograph. But what the Karoo did have in abundance was bones—and, more to the point, teeth. Unlike those who conducted the previous study looking at teeth and bones, Ken proposed to take his samples directly from the rock, at a single locality. In this scheme fossils would be found and their teeth sampled on the spot.

Bones are so porous that minerals replace them soon after deposition, and in this process all their contained organic material is quickly leached out. But teeth are another matter. The thick and impervious carbonate deposits necessary to make teeth function during the life of an organism can trap various isotopes of carbon atoms—including organic carbon—quite intact. Ken was not searching for anything as complex as ancient DNA; he simply needed an accurate assessment of the ratio of two isotopes of carbon from within various levels in the Karoo. And the teeth of the mammal-like reptiles just might give this.

By the time Ken arrived in South Africa in July 1996, the controversy about what caused the Permian extinction had become even more visible in scientific circles. This occurred due to the publication in *Science Magazine* of a startling new theory proposed by a group led by Harvard paleobotantist Andrew Knoll. Knoll had spent most of his career studying environments of the late Precambrian Era, the time about 600 million years ago immediately preceding the advent of large animals and skeletons. To the surprise of the geological fraternity, Knoll found that the latest time intervals of the Precambrian Era were characterized by large-scale isotopic shifts similar to those observed at the later mass-extinction boundaries, such as the Permo/Triassic and Cretaceous/Tertiary. Were these caused by mass extinction, too? The Precambrian Era in earth history was not a time of great biotic diversity, and perhaps even biomass did not match that of the later P/T and K/T boundaries. Knoll and his colleagues proposed a novel explanation. They suggested that the oceans of both the late Precambrian, and the late Permian were "unmixed"—that is, that large amounts of organic material were locked in bottom sediments, but because of sluggish ocean-bottom circulation patterns, little of this material was brought back into shallower surface waters, as it is today in our oceans. Thus, over the last few million years of the Precambrian Era, as well as the last few million years of the Permian, increasing amounts of organic-rich material were taken out of the biosphere and locked into the deep ocean bottom. Then, for reasons still unknown (but probably related to an increase in plate-tectonic activity), this pattern changed. The ocean went from being "unmixed" to being "mixed," and the effect was to liberate vast quantities of carbon dioxide and organic material back into the waters of the sea. At the same time, one of the greatest episodes of volcanism known in earth history also took place in Siberia, releasing more carbon dioxide, this time into the atmosphere. The results were spectacular—and deadly. The initiation of oceanic mixing caused vast volumes of carbon dioxide to be liberated into the sea as a dissolved gas and, ultimately, to be released into the atmosphere itself.

Knoll and his colleagues proposed that the sudden increase in carbon-dioxide content dissolved in the ocean was the killing mechanism causing most marine species to die out. Carbon dioxide in elevated concentrations is a known killer, and marine animals—especially those secreting calcareous shells—are particularly susceptible to CO_2 poisoning. The problem with this model, however, is that it cannot explain the coincident killing of land animals. Most terrestrial creatures are less sensitive to excess carbon dioxide.

It was in this climate of emerging theories and data that Ken MacLeod set out with Roger Smith to collect sufficient samples of tooth material to test these new ideas. They traveled to the Bethel Canyon, where Roger and his crew had made the collections leading to Roger's groundbreaking 1995 paper. Working the slopes, they succeeded in finding and extracting the teeth from eighteen fossil skulls spanning 140 meters of fossil strata. Like a pair of mad dentists from one's worst nightmare, Smith and MacLeod pored over the strata, finding skulls and smashing out the teeth. Because the skulls were relatively few in number, Ken decided to improve his chances at success by collecting nodules of calcite from the strata as well. These nodules had formed in the soil of the ancient floodplain and might also contain an isotopic signal. To make sure that any pattern discovered was not random chance, Ken went to a farm near Graaff-Reinet that had slightly older rocks and sampled nodules there as well.

After a month abroad, Ken returned to Seattle, settled into an office, and proceeded to think about how to analyze his collected samples. The problem was not insignificant. Isotopic signals are fragile things and easily disrupted by heating, groundwater pollution, or any other subsequent geological disturbance. We were asking strata more than 250 million years old to retain a tenuous chemical signature. That signature needed to be teased out of these rocks, and it would take time, trial and error, and many false starts and leads to do so. It took more than a year for the mass spectrographs to spit out data that could in turn be tested for reliability against chemical foul play. Ultimately all Ken's work paid off, and he concluded that a signal was indeed apparent. At the point where the last *Dicynodon* was found, there was a large isotopic spike in both teeth and nodules. The Permian extinction in the Karoo was marked by a single, large isotopic perturbation.

This work added another important piece to the puzzle. It seemingly showed that the extinctions on land and sea took place at the same time—and over a relatively short period. But it also raised new questions to answer: How short is short? How to further resolve exactly how long that might be?

There are no ashes in the Permian-extinction beds that can be dated using radiometric dating like carbon 14 analyses, but there are other ways of telling time in sedimentary rocks. I still had a bit of money left in the grant that had funded Ken MacLeod for a year. I used it to return to the Karoo with a "paleomagnetist," to study Karoo magnetics and try to refine ancient time. But six years had passed since my last visit, and while I thought I was returning to a place I knew well, in reality a new Cape Town awaited a naïve and trusting American.

Chapter 5

Karoo Magnetics

SEPTEMBER 1997

The pack moved in with deceptive speed, loping in a smooth gait until the prey was surrounded and all avenues of escape were cut off. The prey's first reaction was surprise, then a gut-wrenching moment of panic as this new reality became all too apparent; nothing else existed but this moment. The predators were much smaller than the prey, and all were very young, but they were numerous, and that gave them a collective courage. The attack happened very swiftly, but the prey, still in denial, perceived it at a slow-motion pace. The first feint came from the front, a single predator coming in fast, getting the prey's attention, while two more, unseen, dived onto the victim from behind. Another two or three piled on, bringing him quickly down.

I yelled, felt hands tearing at my pockets, other hands grabbing at my watch, and some passive part of me—objectively apart from the pathetic victim now pinned to the ground—waited for the hot electricity of a knife sliding into his viscera. A moment of this passivity, and then blind rage consumed me as one teenager more forcefully tried to rip the watch off my wrist. It was not that the watch had any monetary value. It was simply a Swiss army watch, expendable, but mine, and it had sentimental value. They could have the money but not the watch, or so went the instant balance of decision making, and luckily for me there was no knife, or perhaps they simply chose not to use it; on such ful-crums does fate balance. "No!" I yelled, and some new drive kicked in; I rolled, smashed one in the head with my fist, and was up now, years in the gym repaying their debt. The detached part of me saw my body running toward the nearest street corner, where the normal world of lights and cars could be seen, my long legs now hitting their stride, an electric dose of adrenaline carry-ing me into the clear as the pursuers fell back, and then searing pain as I hit a cable at knee height—meant to keep cars out of the dark park I had so foolishly yet innocently wandered into while walking back to my hotel—and I was down. I rolled and was up again and running once more, ancient chemicals fueling a big-game animal's flight from the hyenas.

But now I was alone, for they had all scattered. The pack that had come on me so quickly and silently and with such clear and cold menace was nowhere to be seen. The predators had slipped back into the dark shadows of the night whence they'd come, to await the next prey perhaps, or to gloat over the five dollars they'd ripped from my pockets—a fortune to these Cape Town boys, children by other cultures' standards. I'd been warned but had believed that I was safe and invincible because of my gender, size, race, and past experiences in this city. But I had come to a new South Africa, a new Cape Town, and crime was now one of the leading employers in the city.

The night was still electric, and the throngs milled, thousands of Africans brought in from the townships in anticipation of Cape

Town's successful bid for the 2004 Summer Olympic Games. All of Africa was doomed to disappointment on this night, for the Olympic Committee played business-as-usual and awarded these games to the Northern Hemisphere, to Athens once more. The announcement was made on a big closed-circuit television in a packed square in Cape Town, and the great party planned for the dizzying acceptance of postapartheid South Africa, a coming-out party in front of the whole world, was instantly crushed with word that Cape Town had finished "a strong second." I had come to see this celebration, to feel the crowds, to take stock of this place I had not visited in more than six years. "Don't go," my colleagues at the South African Museum had urged me earlier in the day, but I would not listen, still living in the Cape Town of 1991, when I could—or at least thought I could—walk anywhere in the city with impunity. Times had changed. To be white meant that one had money and was fair game. The police force had been replaced by "armed-response units" that protected property, not people. These mostly white security units were rented by the month and came in heavily armed and shooting at the advent of any property disturbance. But to be on the street was not one of those property-saving situations. I limped toward a nearby taxi and returned to my hotel, much chastened. A new Africa had welcomed me back.

The next morning I was at the airport, sore, bruised, and shaken, waiting for my partner in this new enterprise, Dr. Joe Kirschvink of Caltech. Joe and I had collaborated on studies of fossiliferous rocks in the region of Baja California for several years prior to this trip, and he brought to the Karoo a specialty that might be of great use in exploring one of the most pressing questions still unknown about the Permian extinction: Was it simultaneous on land and sea, or did the extinction take place at different times in these very different environments? Might the period of mass death have occurred earlier in the ocean than on land, say—or the other way around? The only way to answer these questions was to find the interval of mass extinction on land and sea, and then precisely date the layer of rock that this event occurred in.

Our isotopic work had suggested that the extinction took place at the same time all over the world. But, being perpetual doubters, we wanted to further test that finding with a second dating method of some sort. Unfortunately, dating rock with any precision is a very difficult process, especially fossil-bearing rock. If the rocks being studied contain ancient ash beds, and if the rocks have not been subsequently squeezed or heated or injected with other hot minerals there is a chance. But the Karoo rocks had been squeezed and injected with hot minerals and heated over time—and there were no ashes in any case. Dating them would be a challenge.

One of the most powerful yet least expected geological time-keepers utilizes the earth's magnetic field. Magnets show the curious property of being bipolar—they all have positive and negative poles. Iron filings sprinkled around a magnet will orient in concentric patterns around these poles. Less obvious is the fact that the poles are producing forces that differ so distinctly in direction. Two magnets, when placed together, will either cling to or repel each other, depending on whether their positive or negative poles are in contact. A positive pole will only seek the direction of a negative pole and vice versa. Although the forces emanating from each pole are of equal intensity, something about these forces is markedly different.

The magnets with which we are all familiar, be they horseshoes, rods, or the small disks placed on refrigerators, have stable polarity—that is, their positive and negative poles are always in the same place. But other, more complicated magnets can be made to act in a more variable manner; their polarities can sometimes be traded, so that the positive pole becomes negative and the negative pole positive.

Magnets come in all sizes. The largest we are aware of (although there may be one larger yet out in space) is the earth itself. Our planet's magnetic field emanates from deep within. Its source is the innermost shell of the earth, a liquid iron core, which behaves like a magnetic dynamo. Even though the magnetic field of the earth has two distinct poles, like all magnets, it can be thought of as having been created by a uniformly magnetized sphere. And, like many magnets, the poles of the earth's magnetic field can switch places.

This discovery was made in the early 1900s, when two French geologists, using the most primitive of equipment, discovered that the same lava outcrops in France preserved two diametrically opposing directions of polarity. Since the polarity was detected from rocks long since solidified, these earliest paleomagnetic measurements were looking at fossil compasses, where the directions of the magnetic poles could be ascertained. The fact that two opposing directions were found in the same masses of lava created a debate that continued for decades. Finally, after repeated measurements with ever more sophisticated equipment, there was but one inescapable conclusion: The earth's magnetic field had somehow swapped its polarity, with the positive pole becoming negative and vice versa. The rotation and the orientation of the earth have not changed. The directions of the earth's magnetic field has.

Why reversals take place is a mystery. Many theories abound, and all deal with the complex interactions occurring deep within the earth. The earth's core—the innermost of the three largest shells of earth structure—has never been observed and never can be, Jules Verne fantasies aside. Yet we know a great deal about it, and much of this information comes from the manner and rate at which earthquake waves move through the earth. The core, unlike the overlying mantle, is liquid, but liquid in a hellish way: It is under such high pressure that, although we technically class it as a liquid, it is a type of liquid that cannot exist on the surface. The composition of the core is metallic; it is made up mainly of iron and nickel. Its point of contact with the overlying mantle—which is "solid"—must be one of the more interesting places in the solar system. Complex interactions occur at this core/mantle interface that have enormous ramifications for the rest of the planet. The discovery that the core is made of hot liquid metal showed that it is the originator of the magnetic field, and the perturbations, eddies, convection currents, or other types of movement within the spinning core seem to trigger the magnetic reversals. Perhaps irregularities or gigantic phase interactions between this liquid core and the overlying solid mantle region somehow set off the phase changes. Because we can never directly observe these regions, located thousands of miles beneath our feet, the evidence of how and why a polarity reversal takes place is never direct.

While science pondered the reversals from a theoretical point of view, in the 1960s the implications of this discovery for calibrating geological time became apparent. It was then that geologists began sampling thick piles of lava flows on the edges of volcanoes. Because each individual lava flow could be accurately dated using techniques somewhat similar to radiocarbon (carbon 14) dating, scientists were able to record a relatively precise series of ages for the flows. Each dated flow was then sampled for its *paleomagnetic* direction. To the surprise of the investigators, not only could individual normal and reversed directions be detected, but *many* of these magnetic-field reversals were observable. Geologists soon realized that the present "normal" magnetic-field direction has existed only for about the last half million years. Prior to that the field was "reversed" relative to its present pole directions. As ever older piles of lava were sampled, it was learned that the interval between reversal episodes was irregular but generally quite long, on the order of hundreds of thousands to millions of years.

At about the same time, oceanographers made a similar discovery for underwater volcanoes. In the early 1960s plate-tectonic theory proposed that new oceanic crust (which is composed of lava) is created by long, linear volcanoes arranged along submarine mountain chains called "spreading centers." This new ocean crust is then carried away from the spreading center, piggybacked on a thicker layer of the earth's crust. When oceanographers towed instruments capable of detecting the orientation of the earth's magnetic field over the spreading centers, they observed regions of normal and reversed polarity symmetrically arrayed around the spreading centers. The magnetic signals looked like great striped patterns, which were ultimately mapped across all of the earth's ocean bottoms. The only thing needed to obtain a chronology of reversal history of the earth's magnetic field was an age measurement for each stripe. These were soon obtained by a ship specially designed to sample lava and sediment cores drilled from the bottom of the sea.

By the late 1960s these cores had provided sufficient information about age and polarity that a "polarity timescale" could first be constructed. The greatest advantage of using magnetic reversals as time indicators is that they are worldwide, or "isochronous," time surfaces. By themselves they are virtually useless—there have been so many reversals during earth history that *no* individual reversal is identifiable. However, if combined with other dating techniques, such as biostratigraphy (telling time with fossils) or radiometric dating, the pattern of magnetic-field reversals becomes a very powerful tool. Each time a field reversal takes place, it leaves an indelible mark in the earth's history and provides a worldwide time marker of enormous utility. The record of reversal through time is now well known. That accumulated record is called the Geomagnetic Polarity Timescale, or the GPTS.

▪ ▪ ▪

Detecting the record of these geomagnetic reversals is theoretically simple. Like so many *theoretically* simple things, however, the actual detection process is often less sanguine. The evidence comes from the directions of untold tiny magnetized mineral particles locked within either sediment or lava. The most common of these magnetic minerals, named magnetite, is a rod-shaped crystal with positive and negative poles, like any other magnet we are familiar with. Some of these mineral grains are microscopic in size, and if they exist in a medium where they can freely move (such as in water, or even in unsolidified lava), they act as tiny compass needles. Their positive pole will point toward the negative pole of the earth's great magnet. If present in sufficient quantity, they give their enclosing mother rock a magnetic signal.

Magnetic particles yield useful information about ancient magnetic fields in both volcanic and some sedimentary rocks. The principle of how each of these rock types preserves an actual record of the earth's magnetic-field direction is the same. When hot magma cools and solidifies, the magnetite crystals found within the cooling lava become aligned to the present field direction. A similar process happens when sediments lithify from a wet slurry to solid rock and, in the course of this, lock in place the tiny magnetic minerals, all aligned in one direction by the earth's magnetic field at the time of the rock's formation.

These two types of rock now contain weak—but measurable—magnetic signals. If the exact orientation of the rock is known, a piece of the rock, carefully removed and taken to the lab, can yield not only its magnetic intensities but also the actual directions of the earth's magnetic field when the rock was formed. Thus one can learn whether the rock in question was lithified during a period when the North Pole of the earth coincided with the positive pole of the earth's magnet or vice versa. But to gain that information, a sample or core must be obtained. The rocks need to be drilled.

The instrument necessary to extract the cores is a modified chain saw—a chain saw mutated into a coring device. The chain and saw are gone; the remaining motor is attached to a hollow metal tube about an inch in diameter, coated with diamonds. The motor turns it at high speed, and the thousands of industrial-grade diamonds coating the tip are sufficient to cut several inches into any type of rock. The drill leaves behind a two- or three-inch core, still attached to the solid rock, which is then popped off with a chisel. If the exact position of the core is known (the direction of its long axis and angle of dip into the earth are measured with a compass), it contains locked within it all of the information necessary for decoding the polarity of the earth's magnetic field at the time the rock was formed. Obtaining an oriented core is only the first part of the process necessary to arrive at ancient magnetic directions. Once collected, the core must then be analyzed in a large and complex machine called a magnetometer. The magnetic intensity of the earth's magnetic field is small, and the amount of signal coming from the magnetite grains found within a two-inch-long core one inch in diameter is unbelievably small. The magnetometer was devised to measure these.

It takes two people to drill—one to actually do the drilling, the other to pump water. The drilling requires copious quantities of fresh water constantly pumped through the drill while it is in contact with the rock. The drill has to be muscled into place, and the coring of the rock, accompanied by the constant stream of water being pumped into the hole, creates a fine stream of muddy haze around the operation. The driller is soon covered with wet mud.

The drilling turns out the be the *easiest* part of the operation, for once the holes are bored into the rock, the cores have to be extracted and their orientation noted. Orientation is achieved by inserting a brass sleeve around the core (which is, one hopes, still in the rock, attached at its base, and not now broken off and jammed inside the drill). The sleeve has a platform on top, which is machined so that it holds a geologist's compass. When leveled, the compass gives the orientation of the core, and a second measurement gives the plunge, or the angle of its entrance into the rock. With these two observations, coupled with a measurement of the orientation of the sedimentary beds themselves (which are sometimes in their original, flat-lying orientation but are more often tilted at some angle, a complication that has to be accounted for), a computer can eventually determine the original orientation of the core.

All this takes time—time to select a site, time to drill it, and much time to take the accurate measurements necessary to arrive at core orientation. Because the drill is gas powered, it needs constant refueling and lubrication. The drill bits are prone to breakage. The water can is always in need of refilling.

The magnetic signals frozen in the sediments we needed to sample—or any other rocks, for that matter—maintain their original directions only as long as the rocks were not reheated in the deep past or subjected to a great deal of groundwater passing through their pores. If the rock had been reheated enough, the tiny rods of magnetite that had heretofore so faithfully recorded the ancient magnetic field present at their consolidation into this particular sedimentary rock would reorient themselves to the direction of earth's magnetic field at the time of reheating, and they would do so without leaving any clue.

▪ ▪ ▪

On any research expedition, some sort of vehicle is needed. The romantic view is that a true pickup will be available, or a large four-by-four such as a Land Rover. But the reality is that most expeditions—such as this one—are funded by very small amounts of money, and the most expensive single item is vehicle rental. I had asked for a large-engine sedan of some sort, since the trip would require driving several thousand miles at high speeds to cover the vast distances of the Karoo we needed to traverse. Unfortunately, Avis had no Toyotas left, so they gave me a Daiwa—and a red one at that. I got in, inserted the key, and tried to start it. No dice. Tried again. Silence. I jumped out, went back to the office, and very sheepishly asked why my car wouldn't start. An attendant walked out with me. "Did you use the antitheft key first?" he asked.

"The what?"

I had rented cars all over the world, and this was a first. I discovered that in South Africa it takes two keys to start a car, so high is the rate of carjacking and car theft. Some newer cars are equipped with more lethal ways to discourage carjackers, who jump in at stop signs, throw you out (or shoot you and *then* throw you out), and drive off with your car. I climbed back in, inserted *both* keys, got the thing going, and headed out to pick up Joe and the gear curbside. And so, in this ridiculous car, we set out.

In the six years since I'd last been in South Africa, I had forgotten so much about the place. Especially its size. We faced an eighthour drive just to reach the site of our first sampling area,

Lootsberg Pass, and then we looked at many more hours farther inland to get to our second site, Roger's famous Bethulie site. We drove all day, found a motel in Graaff-Reinet and then set out early the next morning to begin sampling.

The plan was simple. Drill holes, take our small cores from the rocks, fly back to Caltech, and run them in Joe's magnetometer laboratory. Presumably the resultant signal of positive and negative polarities would give us some clue about where we were vis-à-vis the Permian/Triassic boundary, since the polarity signal for this boundary was now worked out from several other places on earth, most significantly from China.

Our first stop was Lootsberg Pass, the high mountain ridge that overlooks a gigantic valley some fifty kilometers from Graaff-Reinet. A new highway had been built over this wall of mountain in the mid-1990s, traversing a relatively low point in the range. The base of the valley stood at about thirty-two hundred feet in elevation, and the top of the pass was nearing fifty-five hundred feet. Somewhere in this pile of rock was the Permian-extinction boundary, but we didn't know exactly where. Things were getting a bit circular. We needed an approximate idea of where the boundary was, but our main objective was to locate the boundary using magnetostratigraphy. Happily, two years earlier the South African Commission on Stratigraphy had published a guide to the stratigraphy of the Karoo and had used the Lootsberg Pass region as its example, or "type section," for the Permian-extinction boundary. This publication had mapped the contact of the lowest Triassic-aged rocks with the highest Permian-aged rocks at a major turn in the road right at the base of the great wall of Lootsberg. We had a starting point.

Finding the appropriate turn in the road that first morning, we began the work in a cold wind. I had somehow gotten it into my mind that Joe and I could drill about a hundred cores each day.

This was way off. On a good day we might do forty, but on some days we could recover only about twenty-five individual cores in a ten- to twelve-hour working period.

The start of things is always exciting because of the novelty, the freshness of a new day, the hope of great discovery. Joe pulled out his drill, attached the water canister that caused a stream of fresh water to pass through the moving drill core, and pulled on the starter cord. The modified chain saw gasped, belched blue smoke, coughed twice, and then came to life with the most horrendous shriek imaginable. Joe was prepared. He had earmuffs to deaden the sound. With this lack of ceremony, we began. I noticed that Joe was dressed in his usual drilling costume, a pair of unbelievably vile pants and an equally nasty shirt. The diamond drill with water pumping through it turned the hard rock it was cutting to a brown mud, which it then splashed through the air. This muddy water soon covered all nearby objects in brown sludge, Joe and me included.

When the first core was finished, Joe pulled the drill out, and we looked at his handiwork. The end of the core stared back at us, a perfectly round circle. Joe carefully extracted the stone core from its hole. It had shattered. He drilled a second, and this one seemed intact. We moved up the hill a yard or so and drilled another and another and another. Soon about ten cores had been drilled, each still in the rock. Joe put away the drill and pulled out a measuring sleeve. This device was a tube of brass that could be placed in the drill hole and would fit snugly over the newly drilled core. It had an accurate compass attached to an arm welded to the long brass tube. The tube was put in place over the core (still resting in the rock), and the angles of dip and plunge of the core were read off the attached compass, these data duly recorded by me. Joe pulled off the sleeve and proceeded to remove the core. Gently he tapped on the core with a chisel. With an audible crack, it popped out, and he slowly removed it from the hole. The core was made up of the red-and-brown shale we had drilled and was about an inch in diameter and two inches long. We marked this first core with indelible ink—LP1—and packed it in its small cloth bag. The whole operation for this core took about fifteen minutes from drilling to putting it safely in the sample bag. One down, hundreds to go.

All through the day the routine continued on the side of this mountain: drill, measure, extract, wrap. We were right in the road cut, and cars swerved by with regularity, some honking, most passing through without editorial comment for the two lunatics drilling holes in the side of a rock. Core after core began to fill our bags as we slowly rose up through time in this pile of strata that made up the Lootsberg mountain. By late afternoon about forty cores were finished. The next day the routine was the same, and we slowly walked up the side of this road, passing through red mudstone and sandstone, the evidence of ancient rivers that existed in the earliest Triassic-aged world of 250 million years ago. Occasionally we even found the remains of the ancient inhabitants of this place, seeing numerous leg bones and the rare skull of the mammal-like reptile *Lystrosaurus*. But these were the odd finds, and I died a thousand deaths from the boredom of this slow work that could yield no clues or results until many months later in the lab at Caltech. At least when one finds fossils, the understanding (and gratification of the new knowledge) is immediate. With the new paleomagnetic game, there was no knowing if the whole procedure would even work.

By our third day we had made it to the top of the highway and still had some time left. I suggested to Joe that we might go down through time from where we had started on the first day, to see what the rocks lower in the section looked like. He assented, and we made our way down a drainage ditch extending under the highway into a meandering creek that skirted the road, and then passed into the wide valley below. To my surprise the rocks here were as red as the rocks we had drilled. How could this be? I was under the impression that the Triassic rocks were red and the underlying Permian rocks green or olive in color or, when cooked, even a rough blue in appearance.

According to our guidebook, the rocks in the creek bed below Lootsberg Pass were Permian in age—yet here they were, bright red in color. Joe loved to drill red rocks, because he said that they gave a much better chance of getting good results due to the iron-rich carrier minerals that yielded the red color. So we went back to the car, grabbed the drills again, and kept drilling, this time drilling downward into ever older rocks.

We eventually found green rocks and drilled these, too. More than a hundred cores had been taken. I also had time to look at these lower rocks, especially at the transition from green to red. The older generation of geologists all had thought that this change happened well before the Permian-extinction boundary. But what if they were wrong?

Joe and I walked this small creek over and over, looking at the nature of the sedimentary strata. Sedimentary rocks are simply the accumulated pile of single depositional episodes known as strata, or beds. By going up, bed by bed, we were looking at increasingly younger rocks. The rocks here began to change color from green to a palette of reds about halfway up the gully fronting the Lootsberg Pass. The green-and-olive strata first showed faint patches of purple, and as we passed successive strata on our journey up through the great stratal column comprising this region, more and more of the red-to-purplish blotches were found within the rocks. We climbed up through an aggregate thickness of about forty feet, and for the last ten feet of this stratal interval, the beds were pure red; they had lost any semblance of green

color. And then a most curious sedimentary phenomenon oc-
cured: One last time, green beds appeared. The most distinctive,
thinly bedded green beds were present in the Lootsberg gully.
These last green beds were very thinly laminated, showing the
finest-scale bedding planes and sedimentary structures. They had
no burrows, no evidence of plant material. They were completely
barren of any evidence of life, with an aggregate thickness of
ten feet.

The strata immediately above and below these thin green laminated beds showed no bedding and were red in color. The lack of
distinct bedding in these underlying and overlying strata comes
from a process known as *bioturbation,* caused when the action of
burrowing organisms such as insects, worms, and crustacea disrupts the original bedding, making it gradually indistinct. Almost
all sedimentary beds are thinly laminated when they are first deposited. But in most environments in our time (and probably in
most of the Permian period as well), the action of burrowing animals disrupts this fine-scale bedding. As years and then centuries
pass, the fine-scale differences in sedimentary composition producing the visible bedding surfaces are destroyed, trampled, in-
gested, homogenized. The resulting rock is massive, featureless,
and free of bedding-plane surfaces. Oddly enough, it is the presence of fine bedding planes that alerts the geologist to the fact
that something extraordinary has certainly happened, for the
presence of such beds indicates that burrowing organisms were
not present. It tells of a world existing in the absence or near absence of animals. And that is rare indeed.

Above this laminated unit, all of the mudstones were brickred
in color. As we approached the road, where our paleomag sampling began on our first day here, we encountered a thick ledge of
sandstone underlain by beds bearing pebbles and bones.

All of this went into our notes, and then we returned to the mindless work of drilling and data recording, the dull drudgery so tedious yet potentially so important. Cores and more cores, drilled, noted, marked, bagged. Each hole was filled with a small metal tag so that in the future we could come back to the same spot and know which hole we were dealing with. These, it turned out, were both effective and hugely important in the years to come.

Joe finally called a halt, satisfied for the moment. He was ready for new vistas, new sections to drill. It was time to move on.

We were traveling now at high speed into new country, places I had never seen, ever inward into Africa. We climbed Lootsberg in our ugly red car, passed through a long plateau of the curious flat-topped kopjes, and eventually slowed to pass through the small town of Middleberg. The next town was Noupoort, and this place offered a distinct shock: a neat part of town for the whites, with tidy houses and white picket fences, and a separate part of town for the blacks, composed of dilapidated shacks and rows of out-houses located some distance from each dwelling. Each "house" was equidistant from its neighbors, and each corresponded to an outhouse also erected with geometric regularity. Scraps of plastic and garbage littered the streets here where the children played. It was a hellish vision, a vision of apartheid, for no normal urban planner would have erected such a place. Several large electric lights stood high over the village and were never extinguished. These were not streetlights—they were spotlights. The place looked very much like a prison compound. Good visibility and clear shooting alleys for police sharpshooters. We were soon to learn that the black neighborhoods were always set up in this way—next to but not within, each small (white) Karoo town.

Several hours into the trip, we crossed the famous Orange River, and in so doing entered the Orange Free State. The land changed yet again, with golden grass replacing some of the lower scrub common in the more southerly parts of the Karoo. There was more game as well, and the beautiful striped springbok became common sights in the fields we sped by. We finally took a road off the freeway, got lost briefly, and then headed onto smaller and smaller roads, finally arriving at our resting spot, a little town known as Bethulie. This was the site of Roger Smith's excavations of 1993 and of Ken MacLeod's work of the year before. But it was still far from the outcrop, and, stashing our gear in the town's only hotel, we headed out to find the now famous canyon.

▪ ▪ ▪

Our first stop was to visit the landowner and secure permission. We found his farm, drove in, and then were confronted by a very large and scary dog who was bent on eating us. There was no way I was getting out of the car to bang on the farmhouse door with this monster circling our car and barking with enough noise to wake the Permian dead. Eventually a boy came out. It was quickly clear that he was developmentally challenged, as I asked over and over to speak to his father and he just stared at me blankly. Joe and I looked at each other, wondering what to do now. We were relieved of any further planning when the farmer showed up, told us how to enter his land, and let us leave.

The gate onto the farm was easy enough, but it soon became obvious that our rental car was not made for this sort of work. We had to drive several miles across a large pasture, where the greatest hazards were the aardvark holes and the termite mounds. The grass was high enough to hide both, and neither of us wanted to rip out the oil pan or break an axle. And so we slowly rolled through this vast field, heading at minimum speed and maximum jolting toward the place where the farmer said we could get access to the canyon. About halfway across I stopped the car, sure that we could pass no farther. We were still a long way from the place the farmer had described as the entry point to the canyon. Eventually we decided to keep going. *What the hell.* I thought. *It's not my car.* Still, if we broke down, the walk back was bound to be awful. Off again we lurched, and I was sure that this adventure would come to no good end. We would roll a bit, hit a termite mound, cut off its top, and then keep going. I began to laugh, putting aside the stress of this drive and how crazy it had made me. We were in Africa, not tied to some desk; the weather was perfect, the countryside breathtakingly beautiful. So what if we cratered out here—a mere nuisance against the larger truth of being able to realize longtime dreams, the childhood years spent reading the books of Roy Chapman Andrews and his fossil hunting in faraway places. Except for the red Daiwa, what more could one ask for? This broad pasture was rimmed on one side by a huge mountain of rock, stacked sandstone and shales rising a thousand feet above our heads. It gave the place a cathedral-like sense, a stateliness, huge monuments to ancient times thrusting up into this clear blue African air. There were no sounds but the sound of this world we had entered into. It was paradise. Not a very good road, but near enough to paradise in all other ways.

We finally came to a stop at the end of the field near a large cattle-watering trough. I jumped out. It was about three-thirty, and the sun was rapidly dropping toward the horizon. There would not be much time for any sort of work, but I really wanted to see the rocks we'd come so far to see, to get a sense of the place before we invaded it, took its measure, made it our own, tamed it. Joe and I scrambled to the edge of the bank and looked in astonishment. The view was incredible. We stood atop a rocky outcrop that gave way into a broad ravine, then into a broader valley, and then into the large Caledon River valley beneath that. Far below, perhaps a thousand feet below, a broad brown river lazily snaked across the terrain. Beyond the river stately buttes rose into the clear African sky, kopjes such as those we had seen throughout the Karoo but immense in size, like the great rocky monuments of Arizona. Beyond them were more high buttes, as far as the eye could see, to the edge of the earth's curvature, a whole planet made up of stone. The panorama was breathtaking—and demoralizing. We had to start on that river and work our way upward to finish our job. The distances were far greater than we had envisioned.

We wanted to see more, so, without much thought, we looked for a way to climb down. The pasture, and the ledge upon which we stood, was composed of welded sandstone, thick and massive, a rock type producing large overhangs that made any sort of descent treacherous, if not impossible. Below us the ravine was manageable, but this top twenty feet or so would require rigorous climbing, both at the start and end of any trip down to the river. Joe and I searched for some way, finally finding a slide with roots that we could slither down. No heroism or dignity here— on your butt, down you go. But we were excited now. The rocks were calling us, and so down we went, working our way past piled rock and shale, under great thorny trees that sometimes blocked our way, expecting, and finding, a much better path at the bottom. It was a huge and wild place of calling birds and laughing monkeys, wild deer and snakes on ledges. Africa indeed.

After an hour's descent, we reluctantly had to call a halt. Darkness was falling, and we still had to get back across the field and drive a half hour to our hotel. I wasn't too keen about driving through this minefield of a pasture in the dark. As expected, it was nerve racking, back wrenching; we were crawling in near dark by the end, finally getting to the gate and the main highway. Joe was cursing as we made our way back to Bethulie in the early African night. It had been a long day, as most travel days are, and we were ready for food. We had assumed that this town would be as lively as Graaff-Reinet, where a good meal could be had at any number of restaurants. Not so in Bethulie.

▪ ▪ ▪

Like all Karoo towns, Bethulie was built around a central and gigantic church, but everything else here was small in scope and size. The hotel was a long one-story building, and we were given keys to our rooms. The place was old, a century old at least, with faded, worn carpet and moldering needlework on the walls. I imagined that entire new families of fungus and mold unknown to science were evolving in its solitude. We had to pass down a long hall to get to our rooms, and there was no light. *The Shining* came to mind, but fortunately it was too dark to see the dead Boers gazing from the numerous empty rooms we passed. Joe and I felt our way through this depressing gloom, and I tripped as the hallway changed levels by six inches with an unannounced step. The rooms were what you'd expect after seeing the rest of the place: swaybacked beds, tiny soft pillows of the kind that leave your neck seriously kinked, a single electric light of about fifteen watts. Huge heavy curtains covered the windows, and the bathroom had an ancient bathtub sitting on four clawed feet. I turned on the water in the bath; a thin tickle of yellow gruel came from the faucet. I decided to forgo a bath and yelled for Joe. He was no more pleased than I. But it beat a tent in the farmer's field—or did it? Too bad for us that we had no tents, no other options than this. How bad could it really be?

We headed back toward the lobby, once more feeling our way in the blackness, trying not to stumble over the raised floor halfway down the dark hall, and went into the ancient bar. Two or three patrons stared morosely into their glasses; the smell of ancient cigarettes and even more ancient beer filled the room like a dark cloud. The barman stared at us warily. "Strangers?" he asked. We only had to open our mouths for our American accents to give that away, and we allowed that we were indeed passing through from other parts.

"What are you doing here?" he asked.

"Digging up fossils," we replied.

"You don't hold with the godless precept of evolution, do you?"

I looked at Joe, looked back at the barman, and ordered a beer.

Dinner was a hearty plate of mystery meat, the staple of the Karoo. Lamb? Kudu? Ostrich? Horse? Baboon? I looked in vain for any sort of vegetables but settled for the meat course. Wise choice! Fasting is not an option after climbing in Karoo canyons. With full bellies we felt better; humans are omnivores, after all, and as long as we didn't know what we'd just eaten, who cared? It would be gone soon enough anyway, the water being as diseased as it was. We went for a walk, covered the town in five minutes, and ended up at the stolid church. What must life be like here? I wondered. The largest store was a video outlet, where movies could be rented to help pass the cold Karoo nights.

At last we returned to our dark, gothic rooms and waited for sleep to come. I sat in my decaying bed thinking about the so-called romance of paleontology, and I finally had to laugh at this circumstance. I had three more nights here.

Morning, as always, improved things. After a hearty meat breakfast, along with much caffeine, we loaded the car for the day to come. We found a small store near the hotel and bought provisions for our lunch. Joe began filling his large jerry cans with water, for the paleomag operation required huge quantities of water, and our reconnoiter of the canyon the day before seemed to suggest that it would be very hard to find. As I was paying for our lunch, the clerk asked me where we would be working. Bethel Canyon, I replied, and the clerk looked at me blankly. "It's Thursday," he opined. Yes, I agreed, it was indeed Thursday. So what? "Thursday is cremation day," he said. Apparently all the bodies accumulating from the week would be burned in the crematorium that day, he explained, then he recommended that we find someplace upwind. Isn't nice to be breathing the dead people. This got a look from me. Bodies? Cremation?

Sure enough, on the way out of town, we passed a tall brick chimney that I had not noticed the night before. The bodies of the many Africans who had died of AIDS that week were being piled up, and a huge fire was being started in the brick fireplace beneath. The pandemic had struck this part of Africa with ferocity, and the death rate was staggering. We jetted out of town with alacrity as a thin gray smoke began to rise up from the chimney.

▪ ▪ ▪

We made our way to the valley once more, again slowly crawling over the large field, trying to avoid the holes and termite mounds. Time and time again, we scraped the bottom of the stolid car so unsuited for this type of fieldwork. This ridiculous Korean car was trying to kill us. Or, more accurately, I guess we were trying to kill it. I wondered then, as I do now, why there isn't some large box printed in red on every rental car form: "I swear on all that is holy that I am not a geologist and will not be using this rental car for geological fieldwork." How many rental cars must the fraternity of geologists have trashed?

We arrived once more at the edge of the great canyon and tumbled out of our vehicular deathtrap. Silent recrimination from Joe: I knew his looks but didn't care. It was time to get to work, and we had far more severe problems. The water had to be carted into the valley.

The next few hours were torture. The sun rose high and hot; the enormous carboys of water had to be manhandled, one by one, down into the canyon. We carried the huge canisters, rested, carried, rested. At least it was downhill. Because we would start our drilling program at the bottom of the canyon and then work our way back up, we had to stash sufficient water to make sure we could drill from bottom to top. Finally, by late morning, we began the long descent into the canyon, trying to find our way to the river.

It was much farther than it seemed when looking from the top. For nearly two hours we made our way down the canyon, sometimes on good trails but more often than not on animal trails and through great thickets of thorn bushes. Joe and I sweltered under an ever hotter African sun with the heavy paleomag gear. We were following a small stream that had cut into the surrounding sedimentary rock, and this canyon ended in a rocky wall a hundred feet or more over our heads. In the upper reaches, the rock was hard sandstone and conglomerate, but as we descended through sometimes tortuous passages the rock began to change and become dominated by more shale than sandstone. The shale was red in color, and bone was abundant. Numerous weathered skulls were loose in the streambed at our feet, and here and there I could see fragments of bone sticking out of the red shale wall.

Eventually the land flattened out, and I knew we must be nearing the river. Large reedy flats and muddy banks appeared, and then the river itself. We lay near the banks on the hard rocky outcrop, resting after this long walk. Now we had to reverse that voyage, going back up, and drilling as we went.

I wanted just to sit on the broad expanse of river for a while, divining, perhaps, what the return voyage and the ensuing work would entail. The Caledon River was good company in laziness. It was wide and muddy, and I imagined that here would be hippos, or at least crocs, sunning in style. But there was no game at all. Perhaps it was the time of day, or perhaps the neat farms now found throughout the region had taken care of any game. What there was, though, and in some abundance, was fossil bone.

The green strata along the bank were from the ancient rivers of 250 million years ago, just before the extinction. Roger Smith had long ago logged the fossil content of this outcrop on the wide Caledon, but he had never collected it. Two large skeletons of *Dicynodon* lay immersed in the green rock, smashed and distorted skulls leering skyward.

I had a copy of the measured section made by Roger Smith some years earlier. It was a tabular column of the strata broken down into meter-thick units, describing each distinct rock unit that we would cross as we walked the long canyon back toward our car. Here on the river bottom, we were late in the Permian; on top, where our car lay, we would be well into the middle part of the Triassic. Ten million years of time lay in between, represented by the rocks making up the canyon, flat sedimentary strata of an ancient time. With both the map of the region and the measured section, we could locate ourselves geographically and stratigraphically and put our cores into a context that would match with Roger's fossil collections.

My job was to figure out where we were in both senses, and Joe would then look for a likely place to drill. Joe was very particular about this. He hated sandstone and loved red shale best of all. Unfortunately, there was almost no red shale in the Permian parts of the section, and because of this, Joe had relatively little hope that any of our efforts in the Permian would work out.

The first drilling was easy, for we were in hard Permian-aged sandstone that took the drill and spit out a beautiful long green core. But Joe was uneasy with these rocks; in his opinion the only rocks worth drilling were the red shales and mudstones. We settled into the routine we'd established over the previous days at Lootsberg: Joe would drill, and he would orient the core; I would record the data, mark the core, wrap it and put it away. By mid-afternoon we had reached the first outcrop of the red mudstones, and an unpleasant shock awaited us. The mudstones would not drill. Joe took first one core, then another, and each time when he pulled the drill out, there was nothing but a pile of red mud and broken pieces to show for it.

With this new development, I received a new job: I was the gluer. Joe would drill in the red mudstone, splashing water and broken bits of rock everywhere, and at the end he would pull out a few shards of mangled rock. Sometimes ten or twelve pieces were all that was left of the core, and I had to figure out their orientation, put them back together, glue them into one piece, and try of make a complete core from the bits. Sometimes it worked, more often it did not. The day was racing by. We were accomplishing very little as core after core shattered.

It was getting to be late afternoon by now; we had fewer than twenty cores to show for an entire day. We stowed the gear under a tree and then started back. The climb up was numbing and hard; we had worked all day already, and a long climb out at the end was not what the doctor ordered. I was dripping with sweat when I crested the last rise, to finally see our red car holding the pasture hostage. Joe climbed over after me, no less beat. We lay there, panting for a while, and then made our way to the car. A whole day. Twenty cores. Not good. Plus, we stank. No avoiding the bath on this night.

Dinner (meat!) and another awful night in the mausoleum-like motel; I was imagining that the gray dust I saw on all the furniture had drifted in from the crematorium. Everything was dusty, and it might just as well have been from the crematorium. I was getting to know Africa; I was breathing Africa.

The next day we resumed where we had left off, drilling again in the red shale. Now the routine was set—and numbing. Joe would drill a core; I would locate its position on the stratigraphic column and map; Joe would extract the pieces; I would spend the next twenty minutes trying to glue them back together. Some could be salvaged; others were useless and had to be jettisoned. We both worried about the numerous dolerites here, for there were more of these invasive igneous rock bodies than at Lootsberg, and we knew that too many of these would doom the project, for too much heat would only reset the cores and destroy any primary information. Unfortunately, there was no way we could know if this would work or not; we could only keep drilling. Again the day passed in tedium.

On this day heat and flies came with a vengeance, and Joe and I had finally had to resort to holding cloths over our faces to keep the flies off. They were relentless, always looking for the eyes, and this pestilence slowed work even further. By the end of the second day of drilling, we had only another twenty cores collected.

We came back a third day and drilled as well, finally calling a halt at midday. We had enough, for we had covered the interval identified by Roger Smith in his 1995 paper as being the Permian-extinction boundary. For good measure we drilled up into the lower Triassic and then hauled all of our gear out. Although it was now midafternoon, we had no stomach for any more nights in the hideous Bethulie Hotel. We loaded up and headed south, back toward Lootsberg, back toward home.

We spent one more day at Lootsberg, again drilling, this time not on the highway but lower in the gully, and my work at Bethulie had now sensitized me to a sequence that seemed to span the Permian-extinction boundary. There was a similarity between the rocks in this lower Lootsberg gully and the transition at Bethulie. A pattern was emerging. We climbed out of the Lootsberg Pass gully this last time, and I bade good-bye to the place, never expecting to see it again. Little did I know that my time in Lootsberg was just beginning, not ending.

Our last day in the field was spent drilling rocks at a place I had never been, in rocks slightly older than those of the Permian-extinction boundary, a place where Ken MacLeod had sampled as well the year before. We wanted to see if these older Permian-aged rocks would give any signal, and, on Roger Smith's advice, we had contacted a farmer north of Graaff-Reinet, at a place called the Brody farm. Mr. Brody had become a hero to all at the South African Museum in the early 1992, when its collecting team was refused permission to camp on the Graaff-Reinet public campground because one of the party was nonwhite. The Brodys heard about this outrage and invited the museum team to camp on their land. Providentially, it turned out that this farm was also one of the richest bone-yielding areas in the Karoo. Skeleton after skeleton was collected and removed from the Brody land as each year Roger Smith took his team back to the site. Joe and I had decided to drill parts of this farm.

Although we were expecting a "farm" in the usual sense of the word, what we found was a wild wilderness of stacked rock and great cactus fields. It was a scene of badlands, not green fields. The Brody family had imported cactus to serve as food for the sheep they raised in times of extreme drought, when the rain truly failed. There had been drought in the Karoo on my first visit in 1991, and here in 1997 I was shocked to learn that the same drought still held sway and that many of the farmers were using the cactus, since all green vegetation had been consumed by the sheep. We had noticed that the dam behind Graaff-Reinet was empty, and Mr. Brody told us of the dire water shortage in the region. Well after well had run dry all through the Karoo, a combination of too many farmers with too many sheep and not enough rainfall.

Mr. Brody took us to the outcrop we wished to sample, a place at the edge of a giant mountain of stacked Permian rock. We had been working now for nearly two weeks without a break, either driving or drilling and sampling each day, and we drilled on this day without a huge sense of purpose. A battery of black monkeys shadowed us and hectored us, and we managed to climb one of the lower peaks for the exercise and the view. But it was clear our hearts were not in this lower sampling, and we called a halt in the afternoon.

On the long walk back to the farmhouse, I spotted a round white object that turned out to be a monkey skull, mostly clean but still having a few patches of rotting flesh on its brow. Normally I would not have stopped, but my weeks in Africa, and all we had done, had made me perceive things somewhat differently. Seeing bodies stacked up and burned, seeing the homeless and heartless, seeing the speed of life and death in this continent had somehow brought a new sense; my views of life and death were slightly altered, and this vervet monkey skull now seemed like a really cool thing to possess. I put it in a plastic sample bag, thinking that this would look great on my mantel at home, or at least it would make a good present for my curio-loving friend Alexis Rockman.

Samples and monkey skull all bagged, we made our good-byes to Mr. Brody and his farm, spent a late night in Graaff-Reinet, dined magnificently, and took the long road back to Cape Town the next day, arriving late at night after an all-day drive. We had two days to kill in Cape Town, so we spent time in the museum talking to Roger Smith and trying to make sense of our sampling. At the end of the day, we unlocked our car to drive back to our hotel, but it was immediately clear that dead-monkey decomposition was going on somewhere in the car. The terrible stench of rotting flesh emanated from the trunk, and there was no doubt about what was causing it. Joe and I disinterred the monkey head, now worried about Ebola and other monkey-borne diseases, and were shocked to see that the monkey still had some relic of his brains visible through a small hole in the skull. "So *that's* what smells so bad!" I exclaimed, trying to make light of this putrefaction. Joe was not amused. Taking back a monkey skull was one thing, but rotting monkey brains might not be a prudent item to declare to U.S. Customs. A solution was needed. A strong solution, at that. Had we been in the Karoo, we could simply have put the skull next to an anthill. But Cape Town was far more ant free than the Karoo, so instead we hiked over to the nearest store and bought a large bottle of bleach and a plastic dish, then put the skull into the bleach. All this was left in the trunk of the car. I can't remember whose stupid decision *that* was, but it was probably mine.

The next day, our last, was taken up with packing and storing all the samples and copying the precious notes. This involved some amount of driving, and it was only after several errands that I remembered the monkey head in the back and panicked. Sure enough, the dish with its putrid bleach had tipped over in the back of the car. I retrieved the head, wrapped it well, and tried to mop up the awful smelly stuff in the back of the car—pools of liquid with fleshy morsels of monkey brain floating gaily about in them. There was nothing to be done. We drove to the airport, dropped the car off, and hightailed it out of the country. A thoroughly trashed red Daiwa with bleach and monkey brains in the trunk was given back to the rental company.

For weeks afterward I waited for the rental-car company to send me a bill for the whole price of the car, but nothing was ever said. However, when I presented the now cleaned monkey skull to my wife as a remembrance of my African adventure, the silent, withering look I got said volumes. The monkey now resides in a deep corner of our basement.

▪ ▪ ▪

In December 1997 Joe called my Seattle office from his lab at Caltech. He had results, he said. He went on to tell me that the Karoo rocks were among the most difficult he'd ever had to analyze—and that he had the proverbial good news/bad news report. Great, I thought. The good news was that there were results from the red rocks—though only the red rocks—drilled at Lootsberg Pass. Joe had finessed some good-looking numbers by heating the rock under controlled conditions in his lab to nearly seven hundred degrees Fahrenheit, and in this fashion removed a persistent "overprint" that the strata had acquired when the Jurassic-aged dolerites had invaded the Karoo country rock. He had a record of both normal and reversed-polarity rocks from the Lootsberg Pass and gully. I immediately went to my literature detailing the known record of polarity change across the Permian-extinction boundary as recovered from rocks in China and Europe. To my surprise, these previous records showed a pattern across the Permian-extinction boundary different from ours. In all previous work, the boundary was placed at the end of a reversed interval. At our study site, we saw just the opposite of that situation. There were three possibilities: Our analyses were spurious, caused by a later overprint; our analyses were correct, but the boundary as placed by the earlier paleontologists was in the wrong place; or—most exciting of all—the extinction happened at different times on land and in the sea. This last would be a bombshell.

So I assumed that I had just received the good news—that we had exciting results. So what was the bad news?

Joe gave me the bad news. *None* of the numerous cores from Bethulie, where we had worked so hard, had yielded interpretable results, nor would they ever. Apparently the amount of heat emanating from the earth when the giant supercontinent of Gondwanaland split asunder, during the Jurassic Period, had cooked different parts of the Karoo to different degrees. The Lootsberg region was itself pretty cooked, as evidenced by the dolerites capping all the hills, but there were far more dolerites in the region of Bethulie and more heat in the region 180 million years ago. That excess heat had reset the magnetic property of the Bethulie rocks to such an extent that not even Joe's expert ministrations could peel away the overprint.

It took me several minutes to digest this news. It was bad, all right. I remembered the days there, the nights in the god-awful hotel, the burning bodies, the long car drive—all that time wasted, with nothing, absolutely nothing, to show for it. And then, being a basically optimistic person (or so I like to think), I shrugged it off. How bad could the news be? We had the results from Lootsberg. So I told Joe that we would simply publish the results from Lootsberg. But he drew me up short. Science is about repeatability. The reason we drilled two sections in the first place was to demonstrate a repeatability of results.

I still didn't get it, so Joe gave me the bottom line.

"We have to go back and drill another section."

At the time I had no plans ever to go back to the Karoo. That changed.

Chapter 6

A Change of Rivers

SEPTEMBER 1998

The large green grassy park was fronted by indigo sea and yellow sand. Rolling white breakers crashed in through giant stands of offshore kelp, causing the kelp's black, vegetative heads to bob in knowing affirmation of the perfect beauty of this place. Cape Town stretched behind us, rising onto the maroon and variegated slopes of Lions Head and Table Mountain. The city center was a bit north, and we enjoyed the early spring and warm sunshine of the crisp oceanfront at our new home, Sea Point. There were swing sets, slides, and teeter-totters. An ice-cream stand off to the side along a busy street filled with cars and a front of perfect hotels and seashore apartments spoke of wealth, success, and modernity all too rare in the desperate economic chaos that is the continent of Africa.

My eighteen-month-old son was trying to learn the intricacies of riding a swing with middling success, but he had already mastered the high-wire act of the slide—and anyway, who was keeping score on such a day? The throng around us was indeed every color of the rainbow: whites like us; a multitude of brown people—many in traditional Muslim costume, others in the shorts and sandals that such perfect beach weather demands; a group of black youths playing soccer nearby; pink and reddened sunbathers in little costume at all. Jesse Jackson would have been proud. Any human should be proud; the picture was of harmony and brotherhood of races. On such little evidence I mentally nominated Cape Town as a model for tolerance and integration for the world. But today I learned that this view was only skin deep.

Sudden screams spattered my South African version of a Norman Rockwell canvas.

Two black kids came hurtling into the playground. One was keening in bloodcurdling fashion; the other was chasing him, or perhaps simply following him. The first one ran past us, turned his head to look behind him, and smashed full speed into the heavy iron leg of the giant municipal slide. The sound of head on metal was sickening. He was down. Screams changed to wrenching moans. He had ripped open his head with the blow, and like all head wounds, it was bleeding profusely, red splashes of blood falling onto the grass beneath the slide stanchion. I got to him fast, bent down, and examined the wound. He looked about eight years old; his eyes were unfocused. Concussion? I glanced up; his friend had already fled, running into traffic, skipping off cars, melting into the city. The crowd that had gathered around us began to melt away as well. I bent back over him. What to do? Using my T-shirt as a bandage to stanch the head wound's bleeding seemed pretty unsanitary. To my eye he needed medical attention and surely stitches at a minimum. I looked around again for professional medical help. A hundred yards away, on the seaside boardwalk, I saw a police services cubicle, and I could see the uniform of a cop inside. I dashed for the cop stand, sure that a first-aid kit and communications of some sort would be available. The cop, white, saw me running up. Quite breathlessly I panted that a kid had suffered a heavily bleeding head wound. I asked if he had a first-aid kit and if he would call for medical help. The cop stared at me and said, "I didn't do it."

I look at him uncomprehendingly, thinking, *Of course you didn't do it, you fool.*

"That kid needs medical attention," I yelled.

"I have witnesses that I didn't do it," the cop stated.

I couldn't believe this. "First aid? A phone?" I pleaded.

The cop was now in my face.

"Those kids are on glue. Stay out of this, foreigner."

With his hand on his riot stick, he stared me down, then turned away.

I ran back to the slide. My wife and baby were still there, but the injured kid had gotten up, and, as I arrived, he took a few unsteady steps and then waded into oncoming traffic in the nearby street, yelling all the while in a language incomprehensible to me. He weaved his lurching way through the busy seaside boulevard and disappeared up the street his companion had taken. There was blood all over the grass where he fell and a red trail marking his passage. It had already attracted flies.

This was our second week here.

It was the time of Mandela.

It had been seven years since my first trip to the Karoo and one year since I'd spent a month there with Joe Kirschvink. I was on sabbatical, and again I had come to South Africa, this time with my wife and young son. International programs of the National Science Foundation had funded this trip, based on the preliminary results from our isotope and paleomagnetic work of the previous two years, giving me a small amount of money to return to the Karoo. It was not that hard a sell, for the Permian extinction was becoming an increasingly interesting and visible scientific question. I had two goals: to drill another section with Joe Kirschvink that would either confirm or refute our intriguing but unpublishable results from Lootsberg and to do the same for the isotopic results by sampling another section that could be used to corroborate (or not) Ken MacLeod's isotopic results.

The South Africa we were now living in was vastly different from the country I'd known seven years previously, and it was changing so fast that even a passing year seemed to signal enormous transformations. Most noticeable was the omnipresence of cell phones. Early in the Mandela years, South Africa made a decision to go cellular and save the cost of hardwiring the country. Now everyone had one, and I realized that I had come from a place where cell phones were a rarity—in 1998, North America was far behind South Africa in this respect. The technology had arrived here faster than the manners on how to use it, however, and restaurants were always filled with the obnoxious noise of cell-phone conversations. Motorists piled up regularly while talking on the phone. This is all too familiar now, but it was a novelty to me then.

Changes were taking place at the South African Museum as well. These were the halcyon days of Nelson Mandela, that saintly ex-con ruling a country still optimistic that its economic power could lift the millions still in poverty into some reasonable standard of living. It was a time when every institution had to look at itself and ask what it did for the good of the country. What did the museum do for South Africa? And why were almost all of its scientists and leaders white?

It was also the time of the TRC—the Truth and Reconciliation Commission—when the sins of past atrocities of the apartheid nightmare were brought to light, when the murder of Steven Biko and so much else was confessed, when a country tried to heal.

All this was part of day-to-day life. Amid it, the Permian extinction was much on my mind, and it was having a great influence on a new book I was then writing, entitled *Rare Earth: Why Complex Life Is Uncommon in the Universe*. By now I was convinced that the mass extinctions that had plagued the earth deep in time had come close to exterminating complex life on this planet. The implication to me was that mass extinctions would likely have this effect on any planet, thus making the evolution and continued existence of complex creatures untenable for long periods of time. The Permian extinction, when our planet nearly had its higher life forms sterilized out of existence, was the major impetus to this new line of reasoning of mine.

Perhaps it was seeing the city around us descend into an abyss of violence that made me so sensitized to the precariousness of life. The murder rate in Cape Town was staggering. In three months we rarely went out after dark. We had learned that safety was to be found with the day, and that the prudent stayed in at night. We cooked and rented movies. I wrote on my laptop. I went to the museum each day, learned more about the Karoo, and waited for the colleagues who would be my collaborators to arrive, collect me, and head out for fieldwork.

The first American visitor to arrive was Joe Kirschvink, who came to Cape Town several weeks after me, with a graduate student assistant in tow. Joe's main purpose in coming involved other things, with a visit to Namibia, but he agreed to stop off for several days with me in the Karoo to drill a section parallel to those we'd made the year before at Lootsberg so that we could test our findings.

Joe's first choice for a rental car was not available at the airport, and he was given an upgrade for the night—to an Audi A4. He and his good student showed up at my apartment in this gaudy fast car, fully intending to turn it over to the rental company the next day for a cheap roller skate of some sort. But Joe was knee-deep in research grants, and, knowing this, I enlisted his assistant to help me strong-arm him into keeping the Audi. We would need the Audi's greater room to house all the gear for three people for a weeklong trip, we whined. Joe was obstinate. But the next morning the rental-car agency still didn't have a replacement. To my delight we headed out in this luxury vehicle. It was no Daiwa. It wasn't even red.

We rocketed to the Karoo in this car. It was heavily powered, and Greg, our assistant, had surely been a reincarnated Formula 1 race-car driver in some previous life. He loved to drive, Joe and I loved to be driven, and this kid kept the pedal to the metal for hours on end. A hundred sixty kilometers per hour for hours—almost a hundred miles an hour, all on the left side of the road, passing countless cars, always on the right. It was mind boggling. How that car must have aged on this trip! Its engine was always screaming, the tachometer redlining. But we *flew*! We cut several hours off my previous best record to the Karoo, arriving in Graaff-Reinet with an afternoon to relax, an evening to get a good dinner, and a whole night of sleep, with coffee and tea to greet us at the crack of dawn, not to mention a huge, wonderful Karoo breakfast. We arrived on the high Lootsberg outcrop early the next morning and determined that the back side of Lootsberg Pass would serve as a parallel section. Then, for three days, we drilled. The work was far more efficient with three instead of two, and we made rapid progress. We took another fifty cores or so, plenty to allow us to know whether our first results were accurate.

It was on day two of this trip that African ticks first entered my consciousness, a topic that would much concern me on subsequent trips. I was not ignorant of the danger posed by ticks; one of my grad students years before had contracted Lyme disease from a tick bite, and he'd been sorely affected for many months. But I'd never seen any in Africa, and somehow I hadn't thought about there being ticks in the Karoo. That misapprehension changed soon enough. On our second day of drilling we were all sitting next to a stand of tall grass, taking a break. I was idly inspecting the ground, watching the armies of ants doing their usual maneuvers, when I spotted a most peculiar-looking bug. It was slightly larger than an ant and of a different shape, but what caught my attention concerning this particular bug were its gait and direction. It seemed to walk so . . . determinedly. There was none of the back-and-forth search pattern of ants, none of that nervous two steps forward one step back of a foraging ant. No, this bug was going in a straight line—straight for my foot! And now it was climbing on my boot! What nerve! I am usually fairly Buddhist about bugs, but this was coming at it pretty close. And then there was another one, doing the same thing. So I grabbed one of the little fellas, whipped out my field magnifying glass, and took a closer look at him.

What confronted me under the magnifying glass was a most repulsive visage. I immediately jumped up and began brushing off my pants. Two more were on me by now, and several more were on their way. In that moment a new emotion was born—hatred—and damn but I now hate ticks! And at that point I thought I only *really* hated scorpions (one had stung me in Baja California with unpleasant results some years prior to this—by odd coincidence when I was also in the company of Joe Kirschvink). But now I had to include ticks high on my repulsive-vermin list. I yelled to Joe and Greg, and they, too, found several of the tiny arthropoden scum-leg monsters around them, attracted to us humans by our heat and carbon-dioxide signatures. Joe, who knows something about everything, began to pontificate on the subject of peeing around ticks—they're particularly drawn to fresh mammalian urine, he opined. I had visions of the little bastards doing the backstroke as they swam upstream toward a really great blood-rich meal. From that moment forward, it was war on ticks, and we had some really watchful moments every time nature called there in the great outdoors that is Africa.

We had good reason to be nervous about the ticks. The ticks in the Karoo carry a disease that makes Lyme disease seem like a minor ailment. Karoo ticks can carry Lhassa fever, and that kills you. Dead.

That was the start of tick patrols. After marching through tall grass or walking under trees, we would now routinely stop and check for the bastards. And there was no mercy when we found one of these god-awful creatures. Death by rockpick was the summary sentence. If that is what sends me to hell, so be it.

Our next day was a trip to a new part of the Lootsberg Valley, a high section called Wapatsberg. We made our way slowly up this pass and selected a beautiful section of rock to drill and sample. Joe and Greg started the drilling routine while I measured the section. The day went by, and about halfway up the road I saw a most peculiar sight: a pair of browridges emerging from a low outcrop beside the road. I bent down and saw that they were bone, beautifully preserved. I grabbed chisel and hammer and carefully began to excavate, revealing an ever larger skull. I dug around it, slowly removing the strata from all sides, and after much work I got a chisel beneath it and gently pried it free. About a foot and a half long, it was obviously a *Lystrosaurus*, but one coming from Permian-, not Triassic-aged rocks. I amazed my two friends, wrapped it up, and eventually shipped it back to Cape Town. I named it Roger. It became one more datum point—but an important one. After all my time in the Karoo, I was finally learning to see bone.

On the same night, our third day in the field, the two of them put me on a bus back to Cape Town, and Joe and Greg headed north toward Namibia. We now had enough cores to serve as a test section for our magnetostratigraphy, but the analyses, as always, had to await Joe's return to California and running these particular cores on his lab machines, the cryogenic magnetometers that decoded the tiny cores into useful (or sometimes useless) numbers.

The bus came through Graaff-Reinet at 3:00 A.M. My ride back to Cape Town was much longer than my ride to the Karoo had been.

▪ ▪ ▪

Two weeks later it was time for a second, and far longer, Karoo sojourn. My wife was understandably nervous about being left alone in Cape Town, for by now crime was becoming epidemic in the South African cities. It had escalated to bombs in restaurants. The local Planet Hollywood was bombed soon after our arrival, maiming several. American symbols were now being targeted in Cape Town, presumably by Muslim extremists, although in this political climate, who knew? Several weeks later another bomb ripped through the largest shopping and tourist area of the city, the Waterfront Complex. I had not experienced indiscriminate bombing of buildings since Seattle, 1969, a summer of bombs. No Weathermen here, though. Just really angry extremists.

This time it was Roger Smith who picked me up one clear morning, and we headed inland. Roger, who lives in Cape Town, told my wife that there was no danger at all—he left his wife alone all the time! With this dubious well-wishing, we headed out. A part of me hated myself and felt horribly guilty. But after an assessment of odds, I determined that Cape Town had to be safer than the venomous Karoo for a child.

Roger and I had plenty of time to talk and to fill in the history of our two lives led in the seven yeas since I'd seen him. He was different, more reserved, more distant, mysterious. How much did he trust me? He had acquired a poker face that I didn't remember from before. I wondered what I looked like to him. Much was different for both of us, but as regarded the Karoo research, the questions were still the same: How rapid was the Permian extinction among the vertebrate fauna? Was it simultaneous with the marine extinction? How catastrophic was it? And of course, as always: What caused it?

Mile after mile rolled by, and with them the small towns I was now becoming familiar with—Beaufort West after six hours of driving, then Graaff-Reinet after eight. We found our small hotel, the Camdeboo Cottages, and got a night's sleep to ready ourselves for the following morning.

We needed to know where in Lootsberg Pass the Permian-extinction boundary really was, and that could be done only with fossils.

We left our cottage after an early breakfast and prepared for the hourlong drive north. I trudged to the car laden with my heavy field gear and took stock of the day, searching for its omens. The sky was a dark gray and barred with rapidly riding clouds. The wind held the premise, if not the promise, of snow. Incongruous, I thought, snow here in this high desert. But the cold was almost visible, and the Karoo wind only amplified its grip. I had not planned on cold, and had only two sweaters.

Like so many of the Karoo towns, the township area of Graaff-Reinet lies just on the outskirts of town. Here, at 7:00 A.M., little alive was stirring, but the wind whipped up the strewn garbage. The contrast to the neatness and order of the white part of town was striking. Graaff-Reinet was a tourist town in some respects, and kept antiseptically neat to keep that image, but the township was another matter. It was as if the people within neither cared that they were being buried in piles of filth nor could do anything about it. The township itself was the usual collection of huts cobbled together made of corrugated roofs, large pieces of tin, a car door or two, and scavenged wood. Each house was a unique assemblage of what twentieth-century junk could bestow. The windblown trash had accumulated next to the fences around the town, looking like drifted snow peed on by dogs, for yellow plastic sacks from the nearby supermarkets and liquor stores predominated.

We jetted away from this horror at high speed, and soon climbed out of the valley that Graaff-Reinet sits in. Ahead of us rose a huge wall of rock, and for the next ten miles we climbed ever higher, mounting what is known as the Great Escarpment. Then it was ridge after ridge, a basin and range system, with each step leaving us at a higher mean altitude. We crested a final ridge and saw in the far distance a mountain barrier marking the limit of the largest valley to date. "Lootsberg," Roger murmured. It seemed inconceivable to me that the Boers of two centuries ago had conquered this pass and the others with their ox-drawn wagons as they trekked across and settled the Karoo.

We drove almost to the base of the high wall of Lootsberg, the place where Joe Kirschvink and I had spent so many hours the previous two Septembers. I was mystified at first by its appearance—the higher regions of the entire basin were somehow whitened—and then realized that its higher elevations were covered with snow. Roger pulled into a widened area on the side of the road and parked the car. Far away a single farmhouse lay isolated in a sea of low brush and grass. Roger grabbed his gear and set out toward it.

Roger Smith is at least a head shorter than I am—my legs are thus far longer—but his gait was such that I almost had to run to stay up with him. He had a peculiar but highly effective stride in the rough veld; the short, thorny bushes and numerous football-size rocks strewn across the ground made walking a hazardous undertaking, but Roger contrived to have each step land nearly vertically, thus fending off rocks or tussocks. He marched this way across the half mile to the farm at great speed. He walked. I jogged to keep up.

And we needed the speed. The temperature was near freezing, but the twenty- to thirty-miles-per-hour wind had pushed the actual temperature, with wind chill, to far below freezing. I walked as fast as I could to stay warm and tried not to fall too far behind Roger, ruing my lack of correct clothing. But this was the desert! This was nearly tropical Africa! Where was the heat?

We finally arrived at the great white house, and it was clear that this place had long been abandoned. The shutters swung freely in the wind, and the doors were bolted. The farmhouse was two stories tall and made of stone; there was an inscription on it: TWEEFONTEIN, DEDICATED APRIL 30, 1863. A graveyard sat behind the main house. It was gothic, an Addams Family scene, but all too real under the gray sky with the bare dead trees around the house, crows in the blackened branches, a lone buzzard far overhead. At the time I had no inkling that some years hence I would live in this same farmyard in a cold tent under this same blowing Lootsberg wind rolling off the pass for weeks at a time, and that this cemetery would become a familiar place.

The farm was only a brief stop. Roger set off again, climbing and jumping over a succession of barbed-wire fences. These were always dangerous barriers to cross; some people preferred going under the wires, or through, but Roger and I simply climbed up two strands, put our foot on the top, and then pushed upward and over, making sure to clear the dangerous upper strand, then flying through the air and landing (we hoped) on level ground, and not on some ankle-snapping boulder. The landing had to be taken carefully, too, as we were laden with heavy packs, and knees can handle only so much pressure on the high jumps and five- to six-foot descents. Every field day involved four or five such jumps at least, sometimes more. It was the most amazing thing—flying geologists, high jumpers, clearing the dangerous top strand. On two occasions I did *not* clear the wires and ended up with ripped jeans (but happily not ripped thighs or worse) as a result. Badges of courage—or foolishness?

We were out of the flats now and beginning to see a gentle upward slope to the land as it towered up the wall of the Lootsberg range. In various low gullies, olive-colored sedimentary rock appeared for the first time, Permian-aged outcrops. Much higher on the hill, we could see bright red beds, hallmark of the Triassic, and we knew that a Permian-extinction boundary lay somewhere in between.

By now it was midmorning, but the grayness overhead had not parted. It was a strange twilight of a day, more like my native Pacific Northwest than the usually bright Africa. We were on the steepening slope, and Roger signaled a halt. We decided to split up and begin serious searching. This was not welcome news, because the long and fast march had kept me warm from my exertions, but now, looking for fossils, I had to go slow, stoop down, look carefully at the rocks around me for signs of fossils, and concentrate on the visual signals that give evidence of a buried skeleton. The cold hit with a vengeance, and my body, covered with a sheen of sweat from the hard march, rapidly cooled. I stuck my hands into thin pockets to warm them as well as I could, but there was no keeping out the invading cold. It spread first to the skin, working on my face and ears, then spread to my internal organs, gripping my spine. "Africa," I kept muttering. "This is Africa. Supposed to be hot." It seemed more like Antarctica.

I could see Roger off in the distance, the clear air seeming to make things closer than they were, much as clear seawater will fool a diver about distances. The sky was still a hard bar of gray iron, and the wind was like a marching band's brass section, blowing lustily. We were both climbing now, moving up on the swell of the Lootsberg range, the giant valley below us expanding in view as we painfully and slowly rose. Walls of rock were visible on all sides—a clear bowl of view ten miles across and more. There were scattered farms across the valley, but few in number, a brown expanse with only slight hints of green. The rocks beneath my feet were still green, but here and there something new began to creep in: traces of red, a magenta tinge to the finer-grained strata, the lightest promise of a different color in the sedimentary sheets underneath my hard boots. The low outcrops were indeed changing as I rose up this great hill, and into higher and higher, thus younger and younger, sedimentary strata.

The light also began to change as I started to shiver with involuntary tremors from the cold. The wind picked up, and I thought I was beginning to hallucinate—white specks came hurtling into me from the valley air, moving in with stately grace and increasing in number. I had never seen such a thing—at least not in Africa. I caught one with my hand, where it quickly melted. It was snow. Snow was falling here in this high African upland, but snow propelled by the wind, snow coming in horizontally rather than vertically, the way snow is supposed to. I looked in wonder and surprise as the snowfall increased but never fell on me; it smashed into me, this perfectly horizontal snowfall that did not fall but flew. I looked off into the clear air, and the thin snow was not filling it, as in a heavy snowstorm, but coming in like millions of tiny, crazy suicide pilots bent on smashing the earth and us two fools out of existence on this day in this cold place with their hell-bent progress. Horizontal snow, the Karoo. Where was I? This dream place, the cold, the distance from home—why was I here? Is this fun? Does hardship make one a better person or a deluded person? *Quit whining. Clear your mind and look for bone.* Simplify the chaos underfoot, remove the modern, take away the living and newly dead plant life, ignore the loose rock, concentrate on the layered rock, omit the cracks and joints, ignore bedding, now look for the regularity of ancient life. Seek pattern, seek regularity, seek bone.

An hour passed, and we were higher, well above any green strata; all the fine-grained strata were brickred in color, and I guessed that we were now into the Triassic. I saw Roger over the next hill and began the trek to join him. Hunger was gnawing away; I was freezing and needed food for some sort of energy or warmth. How could I find a fossil when all my senses were occupied with how cold I felt? My mind and body fought, one demanding clarity, discipline, and simplicity, the other demanding warmth.

I got to Roger, and, as always, he was unperturbed. I asked then if he was cold, and he seemed quite unconcerned. He told me that he'd already found two skulls of *Dicynodon* just below us, so we knew we were close to the Permian-extinction boundary. We found a slight depression in the rocks and sat down for our sandwich lunches. The snow was still coming in on a flat trajectory, and it gave a sense of wonder to the place, a sense of otherworldliness. Roger was eating stolidly and idly picking at the rocks he was sitting on. He shifted his body a bit to find a more comfortable spot, and quite nonchalantly smacked his hammer against a grapefruit-size nodule that he'd been partially sitting on. After several lusty blows, the nodule came free of its stratal grip, and Roger gave it one sharp crack with his hammer. He cried out in triumph. The concretion had split in two, and a wicked set of teeth set in a bony jaw and palate were visible within the rock. A small but beautifully preserved skull was present. He passed on half the nodule to me. The bones were exquisite, the small teeth looking like they'd been part of an animal living only yesterday instead of a creature lying entombed for 250 million years. This ancient ancestor lying a quarter billion years in this strata, in this grave, until a geologist sat on it while eating a peanut-butter sandwich. For that is what he had found: a cynodont, the survivor of the Permian extinction that gave rise to all mammals, including us. Much later the meticulous preparators of the South African Museum would clean this six-inch-long skull and carefully remove it from its hard matrix, and the ensuing fossil would become one of the prizes of the entire collection of the South African Museum, a perfect skull. While *T. rex*es and other dinosaurs could go on public sale, this fossil and its kind, firmly controlled by South Africa and its strict laws, could never reach public sale, and thus it was one of those few objects discovered and coveted by humanity that indeed could be called priceless.

It was a disconcerting find for several reasons. This particular skull seemed far too young to be from this particular stratigraphic position. If the previous workers' interpretation of where the Permian-extinction boundary lay was correct, it should have come from much higher in the strata making up Lootsberg Pass. Perhaps this specimen was really Permian in age, rather than Triassic, after all? But that could not be; this species was too far advanced to be Permian in age. It meant that the position of the Permian-extinction boundary had been misinterpreted by previous workers. And if that were the case, then there lay before us a whole new ball game when it came to the speed and destructiveness of the extinction.

I was electrified by this discovery—and chagrined. I had nearly sat on the thing, too, and I dearly wanted to make a significant find while with Roger, to prove my worth out here. Roger was absolutely inscrutable, but I had the feeling that he held my fossil-finding ability in some contempt, and rightly so. My whole professional career had been spent in marine strata finding marine fossils, I rationalized to myself. And not just marine fossils, but fossils of invertebrates. Here I was in different strata, looking for different types of fossils of an age I had never studied. Nice rationalization. I might buy it, but there was a bottom line: We were on a testing trip to see if a larger project was indeed feasible; we needed results, and I had to pull my weight. I wasn't. Ideas are fine, but this was not a theoretical exercise. Success depended on finding fossil vertebrates, and lots of them.

We split up again as the snow increased and the top of Lootsberg disappeared into cloud. But Roger had given me my orders for the day, and I had tacitly accepted them, the start of our long and sometimes uncomfortable marriage as collaborators. *Stare at the rocks, put away discomfort, search for the regularity in shape that is a signal of life, not dead rock. Don't think— search. Don't daydream—search. Don't worry about the cold— search.*

Two more hours went by. Neither Roger nor I found anything further. The afternoon was gathering darkness around it; Roger called a halt, and we began the long walk back to our vehicle, now several miles away. As we came closer, we saw that another truck had pulled in behind us, blocking us in. Two white men sat in the cab watching us. How long had they been there? They eased out of the cab, rifles resting easily on their arms, barrels pointed somewhere between the ground and our torsos. They waited for us to approach, watching us warily. We pulled off our caps, and the moment they could see our skin color, the tension eased and the guns went to a less ready position. They addressed us in Afrikaans, and Roger knew enough of the language to describe our activities after they asked him why we were here.

Roger switched to English, introduced himself and me, and now, our best friends, they bade us good day and told us to be careful of marauders and watch out for the rustlers who'd been raiding the local cattle and sheep herds. Another white family had recently been slaughtered by bands of blacks, they told us, and the locals were watching for anything out of the ordinary. Two men going for a walk together on such a day was out of the ordinary. That they were white men seemed to excuse any crazy behavior, however.

▪ ▪ ▪

Odd how ideas come about, sometimes odder still where and how a new idea first forms. We have only the foggiest notion of how the brain works at all; what hope is there in understanding the genesis of any idea? But if the *how* is mysterious, the *when* and *where* are often much clearer. One of the starkest, and perhaps most far-reaching, ideas emerging from our Karoo work was born in a pizza parlor—or at least what passes for one—in Graaff-Reinet on that same bitterly cold October night, early springtime in the Karoo, nearing Halloween in the rest of the world.

After our long day of searching at Lootsberg, neither of us was in the mood for a drawn-out dinner in any of the fancier restaurants. Roger knew of a dive several blocks away where we could get pizza—perhaps the only pizza joint in the vast Karoo. So off we went, still muffled in sweaters and extra shirts against the desert chill. The restaurant was on the town's main street, across from the gigantic, fortresslike church that dominates the region and gives grim testament to the stark lives led by the Boers who founded this enclave of white settlers amid the black nations of the eighteenth- and nineteenth-century Karoo. A gritty wind carried torn newspaper and litter; car guards in their orange vests waited for business; a few black gangsters watched from storefronts with predatory intensity. Like anyone with sense, we did not maximize the time it took to get from our car to the restaurant.

The owners of the pizzeria clearly had never heard of Martha Stewart. Bare lightbulbs hanging from the ceiling, warm soft drinks in cans, paper plates and plastic utensils, pizza that was unrelated to anything even vaguely Italian—or vaguely fresh. I glanced at the patrons seated near us and inspected their "special": gluey dough covered by a sprinkling of nearly fossilized vegetables and oozing fatty sausage. I had visions of the chef in back, wielding a large bottle of Elmer's Glue over his dough. But of course we would eat it. After fieldwork you always do.

What made this night a success was the wait. We ordered our pizza and then waited and waited, and, being geologists, with no one else around to be embarrassed in front of, we talked of what mattered, not what would be of interest in polite company. We talked of the rocks.

I came into the restaurant with my field notebook, encased in its ancient leather holder, still strapped to my belt like the gun of any good cowboy. (At least the rock pick was resting on the floor of the car). I had a pencil, too, because I was familiar with the long waits for food that characterized any restaurant in the Karoo. I was absolutely too tired to make any sort of polite small talk, and I knew Roger well enough that silence would be acceptable. By this time we were like an old married couple and could sit in silence comfortably enough. I could write up notes, reflect on the day. But Roger, it turned out, was in a more expansive mood on this night. So I took advantage of his knowledge. This man knows more about Karoo geology than anyone else alive. He is a little-appreciated treasure in a country that needs its scientists but does not cherish them. I wanted to learn from Roger what the ancient Karoo world might have looked like, both be-fore and after the great mass extinction. And thus, over that ex-tended wait and even after our so-called food arrived, Roger spun out images of a place long ago that once teemed with life and had now turned to stone.

The Karoo landscape looks like layers of rock stacked one upon another, some with fossils, most without—gritty, eroded, frac-tured rock, of size, hue, and composition in seemingly random as-sembly. But keen eyes can soon banish the randomness, and order can be seen. The rocks—for the most part, anyway—were de-posited under circumstances that can be witnessed on the surface of the planet today.

Rivers, as they meander across wide valleys, dump their loads of sediment in predictable ways. In large rivers the swiftest flows can push huge boulders about and eventually lodge them in qui-eter waters where they will roll no farther and will become in-corporated into the rock record. Swamps at the edges of these rivers accumulate fine muddy particles, and the occasional flood throws capes of mud across the vegetated surroundings. Forests grow and fall and sometimes turn to coal; alluvial fans build at edges of steep slopes; soils accumulate and become buried, form-ing hard limey nodes in their loamy hearts; fierce storms un-leash flash floods that carry all before them, leaving behind sheets of mud and wrack. Volcanoes explode in far-distant places, bringing plumes of ash to sift out of the sky onto a wide floodplain. All of these environments and more existed in the ancient Karoo world of more than 250 million years ago, and all turned to stone. Roger was an expert on rivers and the rocks they produced; he knew how the sediments in meandering rivers would move and eventually settle, and in what pattern of rock accumulation.

With my notebook out, I asked Roger the simplest question

and unleashed a long monologue that seemed so simple, yet was really the result of decades of research and understanding: What would the Karoo have looked like just prior to the mass extinction? And so he began.

This wide valley would have enclosed a great, languorous, muddy-brown river the size of the Mississippi. A river so large would have been the center of a universe, the earth before Copernicus. Dominating the landscape, worlds within worlds, a universe of swamps and oxbows and microclimates, a juggernaut carving a great serpentine course as it migrated through time and space, a sine wave in a millennial dance. Bristly forests would have lined the banks and levees, with deciduous trees more than fifty feet high in places standing above a rich understory of shrubbery, a riot of vegetation along the crevasse splays, a riverworld not like that of Philip José Farmer and Richard Burton of the famous science-fiction epic, but a riverworld of the mammallike reptiles. The land away from the river would have been far less vegetated, almost dry land. It is the rivers that count here, the place where land life—animal life—would have been most concentrated.

All of that is but rock now. The rivers, the river valley, even the animals and plants that lined those distant shores, now but minerals and strata. As the rivers wandered back and forth across the floodplains in typical river fashion, they accumulated strata. Over the centuries and the millennia, the seasonal cycle of summer drought and springtime flood accumulated in the sheets and stacks of strata we see now. In this we are on relatively firm ground. But the next question is so much harder: What caused the extinction in the Karoo?

Here Roger paused, because there was no ready answer to this, a rhetorical question, really, and the reason we were in the Karoo in the first place. But it's good to keep asking even the obvious. Before he could spell out for me his view, the pizza finally arrived. Orange grease oozed from its surface. Vegetables perhaps as old as the Karoo were scattered desultorily across this moonscape. Who knew what species gave origin to the meat? Some poor kudu or springbok? An aged sheep? A barking dog from next door turned from nuisance to nuance? We ate with distaste, but cells cried out for food and energy. Who cared about the taste?

When things don't make sense, you start from the beginning. "Tell me again, Roger," I asked. "Describe it all. We're missing something, something right in our faces, something confronting us and mocking us because we're just too stupid to see it. The greatest mass extinction in earth's long history does not happen according to the normal rules of the world. Tell me again."

"But you know all of this."

"Tell me."

Roger once again went through the story, as he knew it.

At the end of the Permian, more than 250 million years ago, the Karoo was a flat expanse a thousand miles from the mountains at the coast. It was crossed by large, meandering rivers surrounded by lush vegetation. The late-Permian strata—the layers of rock that were deposited during this period, which Roger and I had been studying for years—are composed of mudstones, greenish to olive in color, and sandstone. The first of these strata, called the Balfour Formation, consists of alternating layers of mudstone and sandstone that are stacked in piles and bear the distinctive marks made by running water. Specifically, they indicate the migration of point bars (bendings in a river channel) in rivers. The rapidly running water made sedimentary structures called cross beds, which were simply big sand dunes and ripples that are frozen in stone.

Overlying the Balfour Formation is a hundred-meter-thick interval of rocks that is also made up of mudstone and sandstone. But there are fewer sandstones here than mudstones, and the way that the existing sandstones are bedded, and their particular appearance, suggests that some change in depositional system took place. This unit is known as the Palinkloof Member of the Balfour Formation. It is given this separate name because of its distinctiveness from both overlying and underlying strata. Near the top of this unit, the mudstones change in color from green or olive to red, and the sandstones lose their distinctive cross bedding. This suggests that the rivers changed from broad, slow, meandering types—like the Mississippi and Columbia—to many anastomosing (branching) smaller streams, called "braided streams," such as those found at the bases of mountains or at the ends of glaciers.

Our research had led us to believe that the extinction took place while these piles of rocks were accumulating, and that it took place because of this change. And this changeover from a valley dominated by large meandering rivers to a landscape of rapidly flowing braided streams brought about a major change in vegetation type as well. The ecosystem was perturbed, which resulted in the extinction of the many species of mammal-like reptiles.

"And what caused the rivers to change so dramatically?" I asked.

"Tectonics," answered Roger. There was a pulse of mountain building along the coast. The newly raised mountains began to erode, and as they did so the sediment filled the river valleys, and created the change in river systems from high to low sinuosity.

"So, the mass extinction was ultimately caused by a tectonic event?"

Roger, over another slice of pizza, affirmed that this must be the case.

But wasn't the mass extinction here related to the mass extinction in the sea? That was certainly the implication from the isotope work. How could a tectonic event taking place at the southern end of Gondwanaland cause a mass extinction that affected the entire globe?

Roger had no answer to this objection. There was no answer but coincidence, and scientists hate that sort of explanation.

Keep probing.

Roger, was the tectonic event that created all this havoc dated by radiometric dating?

Yes, he replied. There was a good series of dates, the most reliable at 247 million years ago, just when the Permian extinction took place . . .

He stared at me. For decades it was thought that the Permian extinction took place less than 250 million years ago. But in 1995 Paul Renne of the renowned Berkeley Geochronometry Laboratory had given a new and more reliable date of 250 million years ago. The Permian-extinction event and the episode of Gondwana mountain building no longer matched.

And with this there was the proverbial *ka-ching, ka-ching*—or whatever brains do as connections are made—and I saw a new vision. The changing rivers did not cause the extinction. It was the other way around. The extinction caused the rivers to change.

"Roger . . . if most of the plants in your late-Permian world were to suddenly die off, what would happen to the rivers?"

Roger looked at me, thought, looked some more. "There would be a sudden pulse of sediment," he replied.

"And the rivers would . . ."

But we both knew the answer. The rivers would go from meandering to braided.

A new paradigm was in place. A new vision of an extinction event even more horrifying than one already thought to be a defining boundary for that word.

I thought about the sandstone found in earliest Triassic rocks, enormous piles of sandstone, and of the time it must have taken to deposit them. Geologists are used to equating thickness of sedimentary rocks with time—the thicker the pile, the more time represented. But—psychologically—we are so bound by our beliefs and understanding of how the world works in normal times. We see rivers in the modern world and assume that rivers in the past worked in exactly the same way, that the rate of sediment accumulation would be similar no matter what the geological epoch. But this reasoning is a straitjacket. During catastrophes things change. During a hurricane or major flood, it is not "business as usual." The Permian extinction—the greatest catastrophe for life in the long history of this planet—was clearly not business as usual. The great piles of strata found above the Permian-extinction boundary could have been deposited *not* in the millions of years that all previous geologists had surmised but perhaps in thousands. A sudden destruction of plant life could do that.

The waitress interrupted our musings. The place was closing down. We paid and wandered out into the dark night, visions of destruction dancing in our heads. The world of Graaf-Reinet had not changed during our meal. A small boy came running up begging for money, and then a pack approached. The older loiterers watched us much as a lion does an approaching gazelle, and it was still bitterly cold. But I was only vaguely aware of this reality, living still in a far earlier reality of a world gone abiotic, a world returning to conditions of an earlier time, the time before land plants. What if the extinction was so catastrophic that it killed off most rooted plants, which in turn caused the very nature of rivers to change? An extinction taking our world back to the Precambrian Era.

Like so many ideas, this one was put away for a while and left to ferment—or rot to ridiculousness—by a stern test of time and colleagues. Many months later, back in Seattle, I ran it by a fellow faculty member in my department, an expert on rivers and the ways vegetation affects watershed. David Montgomery listened keenly to this idea, which did not seem nearly so clever as I ran it out for *him* to think about. To my surprise he was quietly enthusiastic and asked if he could think about it for a bit. The next day he was back with a very interesting graph. From literature sources he had tabulated the percentage of geological formations attributed to river types through time and had broken these data into two categories: those river deposits thought to have been produced by meandering rivers, the type found in the Permian Karoo beds, and those from braided streams, such as those found immediately after the mass extinction in the Karoo. I was astonished. There was no firm evidence for meandering rivers until the 400-million-year-old Silurian Period—the time when land plants first evolved. With the evolution of land plants, the number of braided rivers system steadily declined—until the Triassic, when a huge increase occurred. It looked as if changeover from meandering to braided river systems was not a phenomenon occurring only in the Karoo. David beamed in satisfaction.

Were we correct? The jury is still out. But one thing was clear. If the event was as fast as we envisioned, there would be telltale evidence in the rocks themselves. Back on that October night, while I was still in the Karoo, an idea was born, and that idea changed the way we proceeded from that point on. Like the rivers we studied, our thinking took a new course. It was clear that we needed to see more of the Karoo, to move farther afield from the confined area around Lootsberg Pass. A new mission had been launched—to search for distinctive boundary beds. For the first time, we had a model to test.

■ ■ ■

The morning after our memorable pizza, Roger Smith and I drove back to Lootsberg Pass and walked the lower gully, down where all previous workers had considered that only Permian-aged rocks could exist. Within an hour of beginning our search, Roger had found the skull of a *Lystrosaurus*—index fossil of the Triassic—in these supposedly Permian-aged rocks. It was a demonstration of something that Roger had shown in Bethulie—that the *Lystrosaurus* was really a Permian organism that had survived the Permian extinction. It also demonstrated that the Lootsberg region still had fossils to yield. The years of collecting here by workers coming before us had not cleaned out this region, as so many thought. A systematic study of fossils above and below the Permian-extinction boundary—something not even done for the far more well-known and fossiliferous K/T boundary in North America—could be attempted. It furthered our resolve that a study of the ranges of the mammal-like reptiles could indeed be undertaken.

We climbed out of Lootsberg gully and drove ten miles around the edge of the valley to see another section—one we named Old Lootsberg, as the old road over the pass was located here. We drove onto the large farm and pulled up to the gigantic, beautifully restored Dutch Colonial house where the farmer lived. The farmer's name was Kingwell, and we were to find that the Kingwell family now owned much of the Lootsberg valley and, through brothers and cousins, many other farms across this part of the Karoo as well. Charles Kingwell answered the door, asked who we were, and then, in the most courteous fashion, welcomed us onto his land. He gave us a key that would take us through the locked gate leading to the old pass road, and we slowly drove this rugged road into a fairy world of badlands. Exposed outcrop causes the fossil hunter's spirit to leap with joy; there was no cover of the rocks here, and the topography yielded acres of strata perfectly exposed for searching. Our hearts were really thumping now. We fell over ourselves getting out of the car, gathering gear, and heading out onto the huge expanse of perfect fossil-hunting territory.

My job here was straightforward: Measure the strata and collect rock samples that could be used to run analyses of the isotopic signatures that could teach us so much. Roger had a different agenda. He was in full search mode.

By midafternoon we had traveled far into the badlands, and I was busy collecting and measuring. The work was tedious and exacting. I had to measure the strata with a staff, record the measurements in a notebook, collect the hard nodules where they occurred, label and store them in cloth bags, and keep notes of where they had come from. I was not engaged in collecting, although I very much wanted to find something of significance. Roger was some distance behind me. I came to yet another barbed-wire fence, this one posing some difficulty because it had been stretched across a shallow arroyo cut into the Permian-aged sedimentary rock. I climbed under it, acting like some belly-crawling solder in the First World War, careful not to get cut on the surely tetanus-covered rusty wires. On the other side, the stream cut I was following forked into two branches. I took the left and noticed Roger coming behind me, clambering over the fence and, seeing me going left, taking the right fork. Several minutes passed, and then I jumped as Roger yelped from a position tens of meters to my right. I walked to where he was and gently swore. Roger was bending over a giant bony white skull, partially eroded out of the stream bank. He turned to me, smiled, and said a single word:

"Gorgon."

I came closer and saw the large, saberlike teeth, the smaller incisors. Much of the skull was still encased in sediment, but enough was visible to see the outline of this giant Karoo predator.

Roger covered it up with loose sediment. "No time," he said. "We'll come back for it some other day," and we continued.

That night, our last in Graaf-Reinet, we had much to think about. The project we'd thought of doing so many years before looked to be probably—*probably*—feasible. Roger had a bigger test in mind. He had summoned his entire field crew to meet us at the site of his breakthrough work on the Permian-extinction boundary at Bethulie, the place where Joe Kirschvink and I had spent so much fruitless effort in the year before. I asked if we would be consigned to the awful Bethulie Hotel again, but Roger just chuckled.

"We are going to live in the Stone House," he said enigmatically.

Chapter 7

The Stone House at
Tussen die Riviere

The two trucks turned off the main highway, crossed a bridge over the Caledon River, and passed into the park entrance. TUSSEN DIE RIVIERE, the stone sign read: "between the rivers." Our search for appropriate outcrop had led to an African big-game park.

For this trip Roger Smith had organized his entire paleontological collecting crew to join us. Paul October, a native South African, worked in the museum as a keeper of the planetarium, but his real love was fossil collecting, He was a huge man, a singer in his church, patient, soft-spoken, a moral force. Hedi Stummer was there as well, immigrated to South Africa from Germany, a housewife who tired of housewifery and in middle age announced to her family that she was starting a new career as a fossil finder. Finally there was Georgina Skinner, a beautiful, strong British woman only recently expatriated from her home on Jersey, a free spirit who had literally left her husband-to-be and her tame English life at the altar, escaping to Africa first to film hunting parties, then to do her own hunting for ancient fossils.

We planned a ten-day collecting trip at Roger's Bethulie locality, but instead of camping on the farmer's field above the site, as usual, Roger had elected to try something different. The game wardens had loaned us their own private dwelling—a small house they'd built for personal guests, overlooking the Orange River. It would turn out to be a piece of paradise.

The park was situated just upstream of the confluence of the Caledon and Orange Rivers. Its several thousand acres were thus enclosed on three sides by water and on the fourth by a gigantic fence with several locked gates. As we drove up the duty road trying to find our lodging, it was clear that the park was indeed stocked with game. A herd of springbok, then gemsbok, came into view almost immediately, and soon we had spotted a half dozen species of the African game once so plentiful in these parts and now relegated to the scattered parks such as this one.

After several false turns, we found the house. It was extraordinary. The game wardens had built it out from a natural cliff found in Permian-aged sandstone lining the Orange River. The entire house was made from scavenged parts—used doors and windows, mismatched planks for the floor—and a spacious deck overlooked the river. There was no electricity, but a small generator behind the house could pump water into a tank on the roof, where it was warmed by the sun, and in this way the clever builders had given the house two wonderful conveniences: a real hot shower and a working indoor toilet. There was running water in the kitchen, a propane refrigerator, a propane stove, and a large outdoor barbecue pit. Every window looked out at the meandering river some fifty feet below. The Orange is aptly named, a sluggish and wide river laden with red sediment making its way across this southern African plain on a long voyage to the sea.

Unloading the trucks and setting up the kitchen took consider-
able time, and it was not until late afternoon that we could set out
and at least look at the rocks we'd come so far to see. We all loaded
into one truck, eager to see, to search, to find a prize, eager for
fossils.

We were packed into the pickup truck, Roger driving with me
riding shotgun and Paul, Georgie, and Hedi crammed into the
backseat, their legs jammed against the rear of the front seat. The
dirt road was a moonscape of potholes, and with each we lurched
and smashed into new parts of the truck, or each other, for Roger
was in a hurry finally to see some rocks. We drove away from our
stone house into the shimmering afternoon heat of Africa.

The park was huge, and we were new to it. As the days went by,
it would become home turf, another territory mastered, but that
first day it was still a cipher. We were on a high plateau between the
two rivers, a vast expanse of golden grassland and scattered copses
of trees. I had never been this far north in South Africa, and it was
quite different from the more southern Karoo I knew—not lush,
but more like a savanna than a desert. Game was everywhere, scattering at our approach, the fleet springbok easily keeping pace with
our car as they ran a parallel track. We drove through a herd of
scattering ostriches, perhaps a dozen of the apterous, giant birds
sprinting out, their great legs churning but their heads never moving no matter the speed of their run or the irregularity of the terrain, perfectly stable platforms for their great eyes carried high
above the black-and-white body on that impossibly long and bald
neck. *Dinosaurs,* I thought, *a herd of dinosaurs,* and then came the
incongruous thought that the dinosaurs were still a long-in-the-
future dream when the ancient sediment rocks we would search
were first laid down as sand or gravel those 250 million years ago.

Stacks of rock punctuated the landscape, almost pyramidal
piles of Triassic-aged strata, and at last we reached the edge of the

plateau and headed down into the canyon. The sluggish Caledon River snaked through a wide valley below, its banks lined with outcrops of the strata we had come to study. Roger pointed out a far canyon on the other side of the river and told me it was the Bethel Canyon, the deep gorge that Joe Kirschvink and I had sampled the year before.

We dropped down onto the river level after several switch-backs, the truck fishtailing on the loose gravel. The heat was immense, but we were all excited now. We had one gate to pass through, a bridge to cross, and we would soon after arrive at our first field site. Roger pulled up to the gate, and I jumped out to swing it open for the truck. A large brass padlock confronted me. Locked. I walked slowly back to the truck. Roger was not pleased. The rangers hadn't told him that the gate would be locked. He thought for a minute, then swung the truck onto a narrow road paralleling the river, toward some low cliffs still within the game park. He was not going to be denied at least the thrill of looking at some rock this first afternoon.

There was no road leading to the outcrops, so Roger turned off the track onto the broad field, creeping along so as to not run a wheel into one of the frequent large aardvark burrows dotting the landscape. It was unnerving, for the tussocks of grass hid the burrows, and a broken axle was not anything we wanted to deal with. We arrived, finally, and jumped from the truck, grabbing packs and hammers. A small dry creek headed for the river, lined with strata. The crew cheerfully marched down toward the Caledon and the enormous cliffs lining its shore. I chose to walk in the other direction—my first solitude in days.

The small creek cut into the Permian-aged shale, exposing it. The rocks were ideal for finding fossils. Huge flat sheets of ancient sandstone and shale flanked the creek for several yards and also made up low banks in places. The bright sun was dropping in the sky, casting shadows on the ground as I walked, and I tried to clear my mind of all thought, to turn down the volume of the incessant internal radio, sweep away worries and anxieties and thirst or thoughts of the food I'd so long been without. The trick of all the great fossil hunters is concentration, to tune out the rest, to go into search mode, to quiet the mind, not thinking, but tuned in to pattern recognition. The outcrop is an almost infinite sequence of small cracks, spheres, color, beds, joints, faults, loose cobbles, blowing sand, hummocks of grass and intervening bushes, spiders and webs and branches of weed—all extraneous, all to be ignored. I walked slowly up the creek bed, I rose slightly in elevation and slowly through time as well. I knew that around the next bed or two of the dry watercourse, I would climb across the Permian-extinction boundary sooner or later, probably sooner.

There were bits of bone here and there, and I left piles of rock next to any fragments jutting from strata for Roger to look at the next day. Eventually I quit looking for fossils and decided simply to walk up the creek to see how continuous the exposures were, how promising, and whether I could find the boundary I sought.

Just as at Lootsberg, the bed colors were the clue. After a half hour of slow searching and walking, I noticed the first hints of reddish color in the otherwise olive mudstones, and soon the overlying beds showed distinct reddish blotches, then a real bed of red strata. Trees blocked my view of the watercourse ahead, and as I came around the corner, I stumbled on an unexpected sight. Up until now all of the beds had been either blocky sandstone or piles of mudstones, both categories rather thickly bedded. But ahead I saw a gully more deeply cut than any I'd yet seen, and in it lay the most extraordinary beds. They were the striped beds of Lootsberg and Bethulie once again, composed of inch-thick layers of green and red interlayered with each other.

The setting sun threw angled beams of light into the gully, and it blazed with color. I examined the beds closely. The total strati-graphic thickness was about five to six feet, and they were over-lain by massive mudstones deep in color. As I walked up into these overlying beds I noticed nodules of whitish limestone. Scat-tered in these were bones. Lots of bones. I had come upon a *Lystrosaurus*-bone bed.

I'd seen just such mass occurrences the year before, when Joe Kirschvink and I had drilled the Bethulie gorge. Skulls, limb bones, backbones, even complete and articulated skeletons lay strewn about everywhere. It looked as though no one had ever collected this place. I felt an urge to start grabbing the bones, but I knew they had to be left for another day, so I simply looked at this old graveyard. This Triassic-aged graveyard.

Shadows stole over the canyon, and I began retracing my route back to the truck. I found the rest of the crew already there, buzzing with excitement. The high cliffs lining the Caledon were also packed with bones, but slightly older ones than those I'd found. We piled into the truck and began our half-hour ride back to camp.

Our first night was complicated by unfamiliarity but spiced with novelty. The road back to the river cut through fields and across roads that would soon be recognizable, but on this night were new. We were chased by the setting sun, in these predusk hours, in a torpid afternoon heat. The game animals were slow moving, so we passed closely through the great herds. At last we saw our house perched on the side of the cliff above the Orange River, and it was a weary crew that finally arrived at the place that was to become the best of our camps in the Karoo.

Late-afternoon light was streaming through the windows, and an infinity of dust motes danced in the air of the house. Each of us had a pad and sleeping bag to bring out, and space to find. I co-opted a corner for my gear and my possessions—a space of my own. The stone house was composed of only three rooms: a large central living room that adjoined the kitchen, a smaller back bedroom, and a tiny bathroom. There's so much personal gear involved in any one of these geological expeditions that we threatened to overload the tiny house so by necessity we moved part of it (and us) onto the gigantic deck that ran the length of the house on the river side.

The hardest thing about the end of the day was the need to cook. The tiredness, the red skin, and raw, windburned faces cried for a cold beer and a good rest, but the body needed food even more. We took turns cooking over the next two weeks, each of us coming up with some specialty for an appreciative audience. The taste of food in the Karoo was like the climate in a way: clear, sharp, piercing. Roger was a master on the large outdoor barbecue, and it became obvious that the national obsession with barbecuing—called *brai*—is akin to religion in South Africa.

I quickly learned that Roger was a man of enormous order and predictability. We got to Stone House each night about six. As we climbed out of the car, Roger would formally thank everyone for the day just finished and then announce that dinner would be at eight. It never varied.

And so a pattern was set. Long days in the field, followed by camaraderie, great meals, and night after night of starry skies. These Southern skies were unfamiliar to me—a Milky Way never before seen from my Northern home, the great Magellanic Clouds like looming fog banks, bright planets rising, the beacons of Jupiter, Mars, and Saturn shining steadily in this Southern sky.

Cold mornings followed by warm but pleasant afternoons were the drill now, the most pleasant climate imaginable. But then the weather began to change, as weather will. The cold mornings gradually grew warm, and in the afternoons we broiled in the sun. Finally, after so much anticipation, I learned about heat. I knew it would come sooner or later; the surprise was only that it had taken so long. But there was no avoiding it forever. In early November 1998 the famous African heat found us.

We had finished our work in the game park by this time; we'd collected a section rich in fossils and were ready to move on. Our task now was to collect a stratigraphic section on the Bethulie side of the river, parallel to the region we had just completed. These were the same sedimentary rocks from the Caledon River up into the box canyon that Roger had described in 1995, that Ken MacLeod had sampled in 1996, and that Joe Kirschvink and I had drilled—to no avail—in 1997. Roger laid out a plan for systematically collecting the beds in the canyon by walking parallel to bedding in a slow search so that every square foot of our outcrop would be scanned by at least one set of eyes or, even better, two.

Roger felt the tenor of the morning, the heat already rising and gripping the countryside at breakfast. We were sweating profusely as we left for our field site at about eight-thirty, and we had stocked great provisions of water for all. The back of our truck was filled with gallon bottles.

The trip to the Bethulie site was long: From Stone House we had to travel the length of the game park, pass through the locked gate, then cross the river on the public road. From there we entered two more gates that allowed us to cross a farmer's field, then again two more gates as we slowly rolled along the floodplain. After climbing a huge ridge of dolerite that was like a hundred-foot wall of solid rock, then descending onto the flats once more, we crossed a small creek and ran out of road. We hiked into the canyon from the spot where we parked the car, a quarter mile up the dry creek bed, the stratal walls climbing up on either side of us until we were deep in a carved valley with stacked rock of the late Paleozoic straining upward around us.

When we set out from the car, we each carried a gallon of water. The weight of the water seemed excessive, for here in the canyon we were out of the sun, and in the morning gloom and shadow, the heat didn't seem so bad. It is so easy to underestimate the need for water when you're not thirsty, and I, at least, grumbled a bit at all this water that Roger insisted we carry. It was too heavy, so as we walked into the canyon, we each stashed half our load of water.

The canyon was perhaps a quarter mile wide, with its sides rising five hundred to a thousand feet above us. It was by no means vertical; while the uppermost parts of the canyon were made up of sandstone—the very thick Triassic-aged sands known as Katberg sandstone—the underlying units were composed of interbedded sand and shale. Because of their softer nature, they eroded not as a vertical wall but as a steep yet still-climbable slope, inclined between thirty and sixty degrees. Our job was to walk along the sides of these slopes at a similar altitude, searching the rocks at our feet for any enclosed scraps of bone, then mark that bone with tissue paper for later examination by Roger. In this way a systematic search of the entire canyon from its base to the highest accessible parts would be accomplished. At least that was the plan. What the plan didn't take into account was the fact that the sides of the canyon were also covered with thorny brush and, all too frequently, high barbed-wire fences, for even here the various farmers had partitioned the land in order to run their sheep.

The start of any day is so easy. The first hour is always a joy; a good breakfast yields fuel to spare; the cool of the morning is a pleasure; there's always enthusiasm, because you have hope of a spectacular find. That might be the greatest draw—that the next stretch of rock will show a trace of bone, which on closer examination is seen to be linked to another bone; that a slight amount

A skull of an upper Permian gorgonopsian, the largest of the Paleozoic predators.

The graveyard plot of the Fouche family, Karoo, South Africa. Quite coincidentally this graveyard was placed on the site of a much older mass extinction, the Permian mass extinction. Lootsberg Pass is in the background.

The venerable South African Museum, set in the Company Gardens of Cape Town, with the famous Table Mountain in the background.

Roger Smith (left) and me in the Rubidge Museum, Karoo. The photograph was taken in October 1991.

The legendary Karoo fossil hunter James Kitching (left), with an unidentified student, examining fossil bones in the Karoo.

A South African kopje composed of sedimentary rock deposited during the last time unit of the Permian Period. It probably took 2 to 3 million years for this vast pile of strata to accumulate. The Permian/Triassic, or P/T, boundary is found at the top of the picture.

Joe Kirschvink (in back) with a student assistant, 1998, drilling for paleomagnetics.

Patrick Ward, age two, being menaced by a large mammal-like reptile skull in the South African Museum—an example of paleontological child abuse.

The Permian/Triassic boundary found on the Caledon River. The "laminated beds" shown in this picture were deposited at the height of the extinction itself, in a world bereft of life.

Roger Smith painting protective lacquer on a large skeleton thought to be a Gorgon, 1999.

Roger Smith and Georgina Skinner cooking dinner in front of Tweefontein farmhouse, October 1998.

Paul October arranging fossils on the porch of Tweefontein farmhouse.

Paul October collecting fossils atop a kopje in Karoo, October 1999.

Lootsberg Valley from upper ramparts. The small habitation to the left in the photo is the Tweefontein farmhouse where we camped, showing its utter isolation.

A skeleton of a moderate-size mammal-like reptile, showing limbs and rib cage. The skull was already gone from this site.

A skull of a Lystrosaurus *emerging from rock, 1999. This is how the prize skeletons are found—by recognizing small scraps of bone coming out of the ground.*

Me next to a fossil trackway in Permian strata, 2001.

The footprint of a large Dicynodon *in Permian strata, with the foot of James Kitching for scale.*

A small mountain of Karoo strata located near Graat Reinett, South Africa.

The rampart of Lootsberg Pass,
with Triassic strata in higher elevations and Permian strata below.

of digging will expose a skull, a complete specimen, perhaps the most complete ever found of this species, or a new species never before found by science. Perhaps it might be a Gorgon. Of all of the treasures that make up the fossils of the Karoo, the Gorgons hold the greatest value—monetary value, scientific value, and the intangible value recorded deep in some part of the collector's brain when the identification is made that this is indeed the most sought-after of Karoo fossils.

More often than not, the first hour passes and nothing is found. The second hour comes and goes, and the mind begins to wander, doubts begin to creep in *(I won't find anything today!)*, and it becomes an effort to keep your eyes on the outcrop. And today came a new sensation that had not been a factor in all our previous collecting: thirst.

The day was keeping its morning promise of heat. At 10:00 A.M. the heat was a novelty for a Seattle boy. It was a heat that held no humidity, that dried the skin and eyes. I began drinking the water I carried.

The first drink was heaven—a long draw, then the surprise of how much water that draw had emptied from the big bottle. And a second, and a third, and I felt better but slightly guilty. I would need water for the whole day. Water came to equal guilt. Too much and you were wasting energy carrying its heavy load. Too little and you might die if you were to fall and break a leg in some remote corner away from the rest of the crew. Choices.

By 11:00 A.M. the heat was like a force or a sound, subliminal perhaps, a low bass rumble that was not invasive but always recognizable there in the background. The slight wind only seemed to increase the odd sensation I felt, difficult to pinpoint at first, until with a start I realized what was happening—that water was being sucked out of my body. In this third hour I was halfway up the canyon wall, moving in and out of shade, working hard on the irregular slope, my boots gripping the scree, climbing up and over the sandstone ledges that divided the more promising fossil sites. On other days such work would normally cover me with a sheen of sweat, and I had no illusions as to why that was not now happening. As fast as the sweat would appear on my skin, it was carried away by the heat and wind; molecules of water changed from liquid to gas, evaporating off me, leaving behind salt. I was taking large drafts of water every fifteen minutes now.

At noon I was dragging, hungry, thirsty, and needing energy. I wanted lunch, which was not an option. Roger, the controller of our lives and clocks, the organizer of schedules, the fearless leader (I obviously was cranky by this point), decreed that breakfast was to be at seven-thirty, lunch at one, dinner at eight. You could set a clock by his schedule. I was starved. An hour to go, and I had not yet found a fossil. And another realization: I had not once had to urinate.

Life in the field seems to accentuate everything. Food tastes better, smells are sweeter, and all things physical seem more pronounced. You get a sense that you are a center of consciousness being carried on a physical body and that the two are allied but separate things. And while the brain may not seem any better— quite the opposite if it is hot—the body is in its element. You— the brain, the Command Center in this situation—*really* take account of how your Mobile Unit is doing. And when said MU is not urinating, you take notice—especially if you've drunk a half gallon of water in two hours and haven't once had to check for the females and find a convenient spot under a tree for a quick pee. Your body has had all that water sucked out of it by the heat. Your kidneys are starting to create stones of their own.

No one in sight, alone in the huge valley as far as I can tell, noon, and I was starved and obsessing about not being able to pee. And the heat kept increasing. The forecast for our area was for heat (of course), and 40 to 45 Celsius. How the hell hot is 40 Celsius? I'm a scientist. I should know these things. So I stashed myself under a tree and did the calculations to convert Celsius to Fahrenheit: $40 \times 9 = 360$, divided by 5 is 72, add 32 = 104—or did I screw up somewhere?—104 Fahrenheit, I could take that. But what about 45 Celsius? Let's see . . . 113 Fahrenheit. People in Arizona live in that all summer. No sweat—*really* no sweat, it turned out, for there was a now a stronger wind, creating the opposite of wind chill. What should we name this? The "wind-cook factor"? Wind heat? *More signs of craziness*, I thought.

Under the tree my heat was dropping, and I began to realize how crazy I *had* become over the past thirty minutes, out of my head slightly, thinking about separations of body and head and wind chill and all the rest. I took another huge drink of water. *Get your lazy ass up and get to work*, I told myself. Who's in charge here? I had been looking for three hours. No fossils for all that time.

I climbed the side of the cliff and spotted Paul October in the distance. Lunch must be around there somewhere, so I headed toward Paul and soon spotted Roger, Georgie, and Hedi in the same area. Unlike me, they'd all had success. Roger, of course, had found numerous fossils, and the others a few, all well below the Permian-extinction boundary and all valuable additions to our understanding of the distribution of fossils at the boundary itself. Many of these were relatively close to the boundary. The fossil hunters were pleased with their morning. I had nothing to report. They humored me.

It was now nearing 1:00 P.M., and Roger signaled for the long-awaited lunch. Strangely, I was less hungry now. Since we overshot my body's idea about when it should be fed, my stomach had calmed down and was no longer clamoring for sustenance. Or perhaps it was the heat.

We wormed ourselves under a thorny tree, out of the blistering sun. The lunch site was less than luxurious. It was rocky beneath our tree and, as everywhere in the Karoo, the home to armies of ants, who immediately crawled onto us, giving the occasional nip with their mandibles. I was also wary of ticks in this place and maintained constant vigil for the wily bastards. Lunch, in my case a sandwich and fruit, had been both heated and compressed, rather like the processes of metamorphism in geology. Nevertheless it was consumed with alacrity, by both the ants and me. By this time in my Karoo tenure, I was simply wary of the ants but had come to terms with them. It was the ticks that obviously worried me. I spread my day pack in a vain effort to get a place to stretch out and look over the countryside. The landscape was bleached, awash with colorless light. It was also filled with sounds: the rustling of dry leaves in the tree over our heads, the insect armies, the noise of foraging birds. My companions, however, were mostly silent, recharging. I looked at their faces, one by one. Paul was alert, brown eyes twinkling. Hedi was in her usual silent but good humor. Georgie was going on in a prattle about something, and the sound of her lilting voice with its wonderful British accent was pleasing and cheering. She had the longest brown legs, the most beautiful face, but how could you look at such things on your "sister"? Roger was taciturn, a stern but loving father to his gathered children, except in this case I was older than this patriarch.

Lunch over, we now faced the brunt of the afternoon heat. The land around us seemed to shimmer even more; the difference between sun and shade was overwhelming, but the most noticeable effect was the continual loss of water from our bodies. The air was draining me of liquid at a faster rate even than in the morning. I was on my last bottle of water now, with about three hours to go. More guilt and worry. I put additional sunscreen on my arms and legs and face, put my hat on, and walked out into the sun. Slam. The heat from the direct sun was a force.

We worked the rest of the afternoon in tandem, walking the sedimentary layers. Aside from Roger, Georgie was the hero, finding three skeletons of *Dicynodon*. My contribution to the day's find was several scraps of bone that were immediately dismissed by Roger. But I had seen things of great interest nevertheless and could rationalize my existence here. The same sequence of beds seen on the game-park side was visible here: green colored Permian beds, then the first appearance of reddish coloration amid the green beds, then some distinctive sandstone layers, then the first red bed, then the laminated beds. With growing excitement I realized that the location of the Permian/Triassic boundary picked by Roger in his 1995 paper, based on the last occurrence of *Dicynodon*, and the period of maximum isotopic perturbation found by Ken MacLeod in his collection of teeth and nodules from Bethulie were both consistent with what I was seeing of the base of these laminated beds. What's more, these rocks looked every bit like those I had found in the game park only several days before, and like those I'd seen at Lootsberg as well. In the late-afternoon heat, I walked up several box canyons to places where the sequence of beds was repeated, and I found, time after time, the same laminated beds. This new understanding sustained me through the late afternoon. A piece of the puzzle had been revealed. Such repeatability could not be dismissed. My instincts told me that this would be an important find. These beds came about when the world had nearly died, when life was rare, when even the worms and crawlers of the mud were gone, when the world had returned to its long Precambrian, pre-animal-and-plant state—and then returned from that brink.

Roger was now bullying us to keep looking, for everyone was flagging. He pointed up to a layer of sandstone, about fifty feet above our position on the side of the hill. "Has anyone searched that sand layer?" he queried, and we looked sheepishly at one another. Although it was in the region to be studied, it would require a hard climb, and, based on my vast Karoo experience (ha!), it looked like a most unpromising layer of rock: too coarse to have a hope of fossils. None of us had been up there. Roger gave us his most withering look and began rapidly climbing the slope. He arrived on the ledge and started to walk it, searching as he did. He had been looking for perhaps only five minutes when he hallooed and beckoned us to come up. It was a tough climb in the heat, and I would have been sweating profusely—if my body and the Karoo would let me sweat. Breathing heavily, Paul, Georgie, Hedi, and I arrived up on the ridge. A large tree had grown up through cracks in the massive sandstone layer, and just to one side of the tree, a flat region of sandstone was exposed. Resting in the sandstone, as it had for more than 250 million years, was an enormous skull encased in rock. It was beautifully preserved, with no other bones around it; a skull, pulled off the rest of its skeleton by scavengers, perhaps, a bodiless message from deep time, a wonderful, beautiful fossil. How did Roger do it? He was always managing to conjure up fossils when he needed them, as either critical finds or lessons to his crew. Did he know this fossil was here? Did he find it on some previous trip and use it as a lesson for us? Or did he look at this sandstone layer, recognize that it was of a shape and thickness that denoted a river deposit where large skulls would accumulate back in the past, back when this parched region was not hot desert by cool river valley in the last days of the Permian? Who knew? Roger wasn't telling.

He looked at his crew with contempt and then began making notes, sketching the fossil in place, taking its photograph. The skull was from a *Dicynodon*. Last survivor of the Permian. We were within several meters of the boundary. I asked Roger if we would collect it. He looked at me with his fish-eye stare. "Why?" he asked. "We have plenty of these in the museum." But I wasn't so sure. This creature was from just beneath the boundary. Did it harbor in its bones some clue as to what had happened that some paleontologist in the next century might find invaluable? Why leave it? But I had no vote in such matters, not on this trip anyway. I vowed that that would change.

We spent our last hour walking and looking, but we were too overheated to see, and no finds were made by anyone except Roger. He seemed unaffected by the heat. Finally, near five, Roger called a halt to our day. It was indeed time to leave; I had finished the last of my water sometime before, and only the knowledge that there were fresh (but hot, no doubt!) bottles of water in the truck was keeping me going.

The walk back to the truck was, for me at least, brutal. My mouth was dry, my tongue caked with mucus. It hurt to blink my eyes, my feet were sore, my arms and legs were sunburned in spite of the oceans of sunscreen I'd applied all day, my legs were scratched from thorns, and I was dizzy from the heat. Paul walked at my side. "Hot," he said. All I could do was look at him.

When we got back to the car and its blessed water, I took a pull from a bottle and ended up downing at least a quart in one go. Roger shooed us into the truck and told us we'd have to do better the next day. The ride back was brutal as well. Two in the front seat, three of us squeezed into a backseat built for two. The rough road jammed us one into another as we climbed the track across the farmer's land. It was twenty minutes just to the gate, then another half hour before we got home.

We were a tired group back at Stone House. The heat there was no less, for the windows of our house had magnified the late-afternoon sun. We sat on the deck trying to cool off and looking at the river below, the broad muddy river oozing past. It was not the orange of its name but brown. It should be the Brown River. I longed to walk down and dive into it, but there was work to do. Dinner had to be made. As usual, I was starving, and as usual, I had to wait. Dinner at eight.

After we'd been working for nearly a week, Roger told us that the next day would be a holiday. We would just work the morning and have the afternoon free.

So the next morning we went back to Bethel Canyon, to take up where we'd left off. The heat was back, and it brought flies.

Dealing with flies was a constant struggle in the field—not because they were anything more than a nuisance but because we all wanted to maintain a certain presence, a gravity, a dignity in our long treks across this difficult landscape. To be constantly swatting at flies did not fulfill that goal at all. Yet the alternative was to have the flies land in your eyes or, worse, in your mouth, where invariably they would be swallowed. That was a god-awful fate, to swallow a fly that had just spent the last hour of its life laying eggs in sheep shit or gorging on dead carrion. So I, personally, swatted, but I tried to do it as unobtrusively as possible.

The flies had been bad when I was here with Joe Kirschvink the year before, and they were still bad. Today there was a relentless buzzing around the face by the smaller flies and an active hunt by the larger horse- and deerflies looking for any unprotected skin. The latter of these took out big chunks of flesh, so they were forces to be reckoned with.

But how bad could a half day be, even with the flies? And we were getting smarter—this time we stashed extra water along the path, just in case. But since it was only a half day, water did not seem to be a concern. We set out once again, this time looking higher in the strata, now only at the Permian/Triassic boundary and just above.

We were looking in red strata now; the olive and green of the underlying Permian rock were beneath us on the canyon floor. We walked the steep slopes, searching for bits of bone. Roger went first, the best eyes of all, and he missed little. As I slowly moved across the slope, I could see the telltale piles of toilet tissue used to mark a find from another skeleton located. Time passed, and I began to think about the peculiar life I had assumed. While most of the adult world toiled at their busy jobs, I was immersed in a deadly serious search for fossils, in a countryside that could kill you all too readily with a slip on the rocks, a step on a snake, or even the breakdown of a vehicle. An even more bizarre realization occurred to me, helped in no small part by the heat, ever increasing as the hours rolled on. I had reversed my previous life's physiology. Where so much of my early scientific career was devoted to the sea, and to life in the sea, I had now become the exact negative of what I'd once been. In those days I put scuba tanks full of air on my back to use as I traveled beneath the sea. But here I had become a creature who needed bottles of water to survive. I lived like an airborne diver, with tanks full of water on my back, able to visit this inhospitable desert only with the aid of this water I carried. I felt as if I were breathing the water I drank, that no amount was enough; it was breath and life itself.

While all this nonsensical hogwash rolled through my brain, I was out there walking my path, seeing nothing, not concentrating, my brain slowly cooking, sweat trying to cool me without success, the flies madly trying to get to the moisture in my eyes. God knows how many fossils I had passed over, and who cared anyway? There were too few of the damn things as it was. I had left home for this stinking, fly-infested hole, and goddamn it, it was now two o'clock. What the hell happened to our half day? I stopped for another drink and found myself out of water.

Damn. Think. Where did we stash the water?

I slid down the scree slope and into the dry creek bed below, losing a bit of skin in the process, and started retracing my steps. I had stashed my reserve of water under a bush. It had been a very distinctive bush this morning. They all looked the same now.

I came across Hedi and Georgie, who both laughed at me, gave me a drink, and sent me on my way. Still looking for my bush. I finally found it, took a long drink, and headed back up the slope.

In the far distance, I could see Roger clinging to the side of the canyon wall, his white T-shirt just visible. I slowly made my way to him. By this time it was nearing three-thirty. He said that we could head back along the uppermost part of the canyon, searching as we went.

By now the sun was a weapon, a blunt instrument; we walked on rock, light-colored sandstone that reflected the hot sun back onto the underside of our bodies. The sun searched for gaps in our clothes to find skin to burn, but most of all it worked on our heads. My sun cap and the dark-haired head within were like a car whose air conditioner is overworked; more heat was being absorbed by my brain than could be dissipated. I now walked in some odd place, like someone awaking from too long a nap. I knew who I was and where I was, but only in a very detached way. I walked along the highest of the shales, beneath a fifty-foot-tall wall of sheer vertical sandstone face, and the shale glinted bright red in the sun, giving this strange world a roseate glow. A fence of barbed wire to clamber over—careful with the legs, swing over and jump, survived another, strung even this high, halfway up this great canyon, suspended between land and air now, feet occasionally slipping. I came to a fork where I could climb even higher to another shale outcrop perhaps twenty feet farther up the wall, and why not? So up I scrambled, to find a great pile of bones. Funny-looking, too, not the usual *Lystrosaurus*, bigger. Then look, there was a skull. Marked it with toilet paper, damn strange-looking thing here in this red strata. Time to go, and I slid downward, found a tree, got beneath in the only shade around, where the crew found me sometime later—what's the past tense of sunstroke? Sunstruck?

Check out up there, I told Roger, and he did. He came back after a bit madder than hell.

"Peter, you lost the nose of that *Proterosuchus*."

"The what?" I asked.

"*Proterosuchus*. You know, first ancestor of the dinosaurs. Probably the oldest ever found. Too bad you screwed it up by losing the nose piece. Let's go. We can knock off early today."

By this time all the crew had assembled. We looked at our watches: 4:30 P.M. So much for our day off. I was now thoroughly two people. As the organizer of this whole enterprise, I was happy to get more data. Who needed days off? As a lowly field-worker, I resented losing some precious time off the rocks.

The *Proterosuchus* was indeed the oldest of its kind ever found. It was to be my most important find of that field season. It wasn't fair. I wanted to enjoy such a find, instead of barely being able to remember it.

My sense of happiness (or lack thereof) always comes with a time delay. This is a curse. I never know how happy I am (or was) until after the fact. As I look back on the two weeks spent at Stone House, I realize that they were among the happiest times of my life. A shared mission, a place of breathtaking beauty, even the privations all added up to a meal with exquisite taste; there was nothing bland about Stone House and our time there.

One afternoon we went swimming in the Orange River below our house. I tried to keep any of that undoubtedly diseased water out of my mouth and never submerged my head, but after the heat of the day, the feel of that cool water against my parched skin was absolutely intoxicating. Roger and Georgie were in the water with me, and the languid current of the river carried us slowly downstream, past the mudflats lining the shore, past copses of willow trees and near-shore vegetation, until we climbed out, ran back up the mudflats, and jumped in again. After a time Georgie pulled off her T-shirt; she was a tall woman with a movie-star body, not shy on this river edge. In the end the mud was too strong a temptation; I forget who tackled whom first, but soon the three of us were covered in gleaming brown, running and sliding in the slick mud lining the shores. "Displacement activity," the psychologists would have called it. Three healthy young people a long time away from their mates, finally going in for a rinse. We found the mud in our swimsuits—and more intimate nooks and body creases—for days to come. We couldn't put our boots back on for the climb uphill to the house, for our feet were too muddy: Roger carried Georgie on his back, and I tried to do it in bare feet. By the top of the hill my feet were filled with small burrs that festered in the night. The next morning Hedi pulled them out with tweezers, teasing me for being a sissy, but my feet were so swollen I could not get my boots on. I did the day in tennis shoes, in discomfort, while my friends teased me unmercifully.

Nights were no less spectacular. Roger, Georgie, Hedi, and I slept out on the deck, and I would awaken several times each night to see a new panorama of stars overhead. The Milky Way here was vivid—the stars piercingly bright and no less compelling for their strangeness. I had grown up an amateur astronomer; I knew the nighttime constellations of the North, and all the stars were like friends. But here there was only disorientation; new constellations and the few old friends still visible were upside down to my senses. It made them even more beautiful in a way. In the dark night, with my small binoculars for amplification, I could see the wonders in a sky hundreds of miles from any city, a sky of absolute blackness. With my companions sleeping next to me and the sounds of Africa all around—there could be few better moments in a life.

Mornings we awoke at first light, and Roger brought us all tea or coffee in our sleeping bags. I can remember turning to look at Georgie and Hedi beside me and reveling the pleasure of waking next to such strong women. Georgie used to tease me about how I pronounced "bath" and "tomato," and we'd laugh. A golden sun rose up over the river, over the far-distant peaks, over the wildness of Africa. There was not a hint of humanity in the view. We five humans living here could have been the last alive on earth. We were completely self-contained, with food, water, companionship, and a mission. The greatest stress was finding clean socks for the coming day.

One day Roger decided to give us another kind of holiday—he let us loose in the *Lystrosaurus* abundance zone.

For reasons still unfathomable, the lowest Triassic strata above the mass-extinction boundary are composed of red beds packed with fossils. Almost all belong to a single species of mammal-like reptile—*Lystrosaurus*. After the hard work of finding the very rare fossils in the youngest Permian beds, the fossils in these oldest Triassic beds were indeed a holiday. There is an irony and mystery to this. These beds were deposited soon after the mass extinction. Yet the fossils—at least of this one species—are very common. Perhaps the extinction caused conditions allowing fossilization to improve. It's not that there were more animals, just conditions that allowed more fossils to form. Or perhaps there *were* more animals—at least of this one single species. They're like the sheep in the Karoo today—about the same size and found everywhere. Our eyes were now so attuned to seeing bone that this abundance of bone was like sensory overload. But it was delightful. We did not keep all the bones we found; we could not, for there were too many. It was a bit like catching and releasing fish with barbless hooks.

The area we were searching was surrounded by tall grass; we

were like children playing, so common were the fossils. It was a holiday indeed, and I forgot where I was. I was walking to a low outcrop after passing through the tall vegetation, all thoughts on the fossils, and I looked down on my leg. Perhaps I'd felt some new pressure. I froze. The fat, bloated, blood-filled body of a tick was protruding from my upper thigh, just below the cuff of my shorts. I walked slowly to Roger, pointed to the tick. He looked closely, took out his knife (which really made me nervous), put it away, and very gently with forefinger and thumb began to squeeze the skin on either side of the tick. After some minutes of this, the tick took its head out. Roger brushed it off and killed it with his heel. Nothing was said. For weeks afterward I kept waiting for symptoms of African tick fever to hit. They never did.

Our final morning, a field trip that seemed to last forever was reduced to a single day in the end. We went through the usual routine—breakfast in Stone House, the gathering of gear, making lunch, bottles of water put in the car—and loaded in. We crossed the vast interior of the game park, heading for the far gate. The plan this day was to look beyond Bethel Canyon, see if there were unexplored side canyons that could be linked to our measured sections and thus allow us more fossils on our range charts. But about halfway to the edge of the game park, someone asked Roger who had the key. We needed a key to get through the locked gate.

Roger fixed us with a dreadful glare in the mirror. "Who has the key?"

There was no key.

Roger let loose with his most potent curse. Not the "damn," not the "shit," not even the "bloody hell." This one was reserved for very bad moments. It wasn't even original.

"Fuck," he uttered in an angry tone.

Louder now.

"Fuck!"

A third time, with real venom.

"Fuck!"

He threw the car into a sharp turn of the dirt road, smashing into the veld and potholes at high speed, and headed back toward Stone House, now twenty minutes away. When all was said and done, an hour had been lost. Roger was livid. Who knew whose responsibility it was this morning to remember the key? It didn't matter.

Because of the lost time, and the late start, Roger changed plans. He decided that we would look at a few scrappy outcrops found in the farmer's cow field just off the road on the turn-in to our Bethel Canyon site.

We piled out of the truck, thankful not to have to take the rutted track into Bethel on this last full field day. The field had a few scattered patches of outcrop, but nothing substantial, and it looked definitely unpromising in my lofty view. But if I had learned anything on this trip, it was to not second-guess Roger, especially when it came to sites of study.

The pasture sloped upward, and in the distance cliffs of sedimentary rock, equivalent of those of the Bethel Canyon, rose out of the tussocks. I walked that long pasture until these rocks were in close view, and I could immediately recognize the succession of Permian to Triassic that I'd seen across the Karoo on this trip. It was clear that the pasture, and any fossils it might hold, lay in the uppermost Permian—not Triassic. The actual extinction layers lay in the sandstone I was looking at, not lower down in the field. Anything we found among the piles of cow bones and cow shit, dried thorny weeds, and an infinite number of flies would therefore be Permian.

Most of the pasture was dried grass, but here and there small rivulets created in the brief and now long-ago rainy season had indeed exposed the characteristic olive-colored sandstone and shale of the Permian rocks. But, to my great surprise, the small areas of outcrop immediately showed bone fragments. As I walked through the tall brown grass, jumping from outcrop to outcrop, I came upon skeleton after skeleton of large dicyn-odonts. None were worth collecting, for all had been distorted by heating and pressure in the long millennia since the Permian. But all were identifiable as *Dicynodon laceriteps* itself, the name giver of the last time interval of the Permian period. All thus become useful data, but the bones rest still in that field.

What crappy bones! Another cruel trick, this dead head I had encountered. It was nearing ten in the morning now, and the heat was rising. I was halfway between the high cliffs and the fence to the road where our truck lay parked, and in the distance I could see the rest of the crew fanned out. I pulled out my binoculars and peered at my compatriots. Hedi was patiently looking in her usual meticulous way, and Paul seemed to be hammering on some distant lump of rock. But when I turned the binoculars in the direction of Roger, I knew immediately that he had made a find. Like an angler with a trophy fish on the line, Roger was hammering fervently but carefully on a large sheet of shale near one of the fences. I could see Georgie feverishly sweeping rock fragments away from Roger's excavation site, and, like any spec-tator watching some expert fisherman reeling in a true beauty, I decided to walk back and watch the show.

As I came closer, I knew simply by watching Roger work that he really *had* made a find. With Roger this was never a surprise—he never came home empty-handed, and he could find bone in granite if need be. But today he looked like an experienced hunt-ing dog that has treed a major prize. All his body language spoke excitement. He and Georgie had uncovered several square yards of loose overburden and soft shale, and I could see that Roger was working on a skull. He didn't glance up when I arrived, just spoke a soft greeting and kept digging furiously around the large skull. I asked him what it was. He looked up finally, and with a mad gleam in his eye shouted one word:

"Gorgon!"

All further prospecting ceased as the rest of the crew filtered our way, drawn by the excitement. Soon we were all removing rock from an ever-enlarging skeleton. First the head region, then neck vertebrae, then a shoulder girdle came into view, then the whole of the backbone and the hind legs. The skeleton seemed complete. And it was large.

It was now early afternoon, and Roger had seen enough. This skeleton needed to be covered with plaster of paris and burlap before it could be removed, so as to not damage the bones. That operation would take a week, and this was our last day. Roger was, for once, visibly excited. At the time I didn't know enough to be excited. I didn't know, as Roger did, that there had never been a complete gorgonopsian skeleton of this size ever collected, that this particular find promised to yield, for the first time, priceless clues about behavior, posture, and lifestyle of this most fabulous and storied of extinct Karoo animals.

Roger was beside himself. The find of this field season, the find of a lifetime, perhaps—and we didn't have time to collect it. There was no question that this particular skeleton would require care. It would have to be left here. But how? It was now so exposed that any passerby would immediately recognize it for what it was—a prehistoric treasure. The out-of-the-way location and the real rarity of amateur paleontologists in the Karoo worked in Roger's favor, but he could not in good conscience simply leave it as it now was, semiexposed to the blistering sun, wind, and occasional rain. The specimen had to be reburied.

I'm not sure who came up with the idea, but it was brilliant.

We covered the site of the bones with big rocks and erected a tombstone at the head of the barrow. The site now looked like a human grave. In this day of AIDS, when about a third of adult South Africans were infected with the disease, scores dying each day, no one tampered with fresh human graves. As the day came to a close, we left the site of the find with confidence that it would lie undisturbed until Roger returned on the next expedition, a date as yet undecided.

On the way back home, Roger was berating the gods for leading him to a prize skeleton on the last day of the field trip. Soon it was too much—I had to remind him about the circumstances of *why* we were *there,* in that particular location in the first place, and how forgotten keys can metamorphose into found treasure. That earned me his famous dead-fish stare.

Time is so peculiar in its way. There's usually too little or too much of it. Now there was far too little. We were on our last day, and I had to know: Were the boundary beds so obvious and observable at Bethulie and Tussen die Riviere simply an aberration of the northern region, or were they something more universal? If the latter, the thin, laminated bed that Roger and I now suspected to be the boundary beds would be found everywhere in the Lootsberg region as well, as I thought that they were. But I needed to reconfirm that. Memory is tricky. I needed to see three places in quick succession on our way back: Lootsberg Pass and its environs, with its three sites, known as Lootsberg, Old Lootsberg, and Wapatsberg.

After the long and intense heat we'd encountered at Stone House, we all assumed that hot weather would follow us home. It was, after all, late November now, late spring by Southern Hemisphere standards, and midsummer by Karoo standards. So, assuming the best, why were we surprised at the worst?

On the three-hour drive from the Bethulie region toward Lootsberg, the sky became ever more overcast, and by Middleberg we hit a first few thin squalls of rain. As we climbed up the east slope toward Lootsberg Pass, the rain became thicker, and we could see that the pass itself was obscured in cloud. As we rose onto the summit, we hurtled into a dark night. A thick fog gripped the top of Lootsberg, and Roger had to stop and pull over. It was afternoon in a place that is, more often than not, baked in sunshine, but now we were in a dense London pea soup. Huge trucks would loom out of the fog, hurtle by, and be swallowed up. I needed to see the rocks. I could barely see my hand in front of me. Time. Never enough, especially when one takes weather for granted.

Roger slowly wheeled down the pass, parking near the entrance to the lower Lootsberg gully. He intended to take Paul and Hedi and plaster the small carnivore skeleton we'd found on our last trip here, a skeleton still cached in surrounding rock. He gave me the keys to the second car and set me free. I had an afternoon to drive to the three sides of the valley and check the various boundary beds. Georgie jumped in; she was not needed for plastering the small carnivore and had asked to go.

Our first site was nearby. We drove to the base of Lootsberg, where the highway makes its great loop from the straightness of the valley floor onto the rise of the pass itself. We parked beside the road, cars occasionally coming out of the dark gloom, whizzing by, and immediately disappearing into the fog. It was cool, impossible to believe that only the day before we'd been in the absolute furnace of an African desert in the hottest season. Now we could have been in San Francisco, in an impenetrable gray mist, our rain slickers glistening. I scrambled over the guardrail and headed down into the Lootsberg gully. I'd been there so many times before that I knew where I was going. And I was still curious—would I find the boundary? Would I see the thin laminated beds so clearly demarcated at both the game park and the Bethel Canyon? Would I see the telltale signs of changing rivers, the lack of trace fossils and bioturbation, the hallmarks of disaster?

I arrived at the place where I thought these beds should be, but in the darkness and fog my landmarks were virtually invisible. So I looked for any sign of our previous work there and found to my relief that the markers placed in the paleomag drill holes so patiently excavated by Joe Kirschvink the previous year were still in place. Each hole had its small aluminum tag resting within. I found one of our tags and tried to walk through the succession of beds. Now I started to see signposts of lithology. The pattern made sense; it was the same from Bethulie. Two more places to go.

We drove through the fog into the valley and turned up toward Wapatsberg Pass. The road was a dirt track, and we proceeded with caution on this terrible road, in this fog. I finally arrived at the high pass and walked the outcrop here. Same pattern as the other two places.

It was getting late in the afternoon; we had one more stop. I drove with Georgie up to the last pass of the great valley, a place on a farm called Old Lootsberg Road. Here, too, we jumped out to see the same succession of rocks. My hunch was right—the boundary beds representing the Permian extinction were all the same, at the same place in every stratigraphic section. I was elated. Nature had written us a message in the rocks: "At this place and time, one world ended in extinction, and another began." The day was nearly over; the gloom had thickened, and we needed to meet our friends in Graaff-Reinet for dinner.

Dinner was a celebration at the fanciest restaurant in Graaff-Reinet, in the famous old Drostdy Hotel. It was gourmet quality, the wine flowed, but there was sadness, for this trip was over, and soon we would again disperse to our other lives.

But there was also a promise for more. The Gorgon was awaiting collection from its stratal grave. It demanded our return. No one objected.

Chapter 8

Retrieval

Pasadena, California, in the depth of what it calls winter, is breezy, sunny, and crowded. The fresh clear air illuminates the San Gabriel Mountains at its back, clear in this antithesis of the perceived Los Angeles–area atmosphere. I had returned from South Africa in December and had traveled to Pasadena several weeks later to help Joe Kirschvink analyze the cores that we had collected from Lootsberg Pass.

My job in his laboratory at Caltech was simple: I was gofer and lackey, whose main purpose was to physically prepare cores so that they could be placed into the magnetometer and analyzed. This involved taking the cores out of their wrappings, trimming them down to a standard size with a saw, and then polishing the ends to remove any irregularities. Once of an acceptable size, the cores have to be painted with a temperature-proof paint; in the oven in which they will be repeatedly heated, the normal ink we have used in the field to give them their identifying numbers will be baked off.

All of this takes time. I would much rather have been in a situation where I could have done something more heroic—like running them through the magnetometer or recording the all-important data as they were analyzed—but Joe was doing that. So into the windowless, poorly ventilated, and incredibly dirty rock-trimming room I went, to spend my days in this endeavor.

A magnetometer is an odd device. It is a strange, bulky thing, cooled by liquid helium. The small cores we had extracted from the rocks of the Karoo were sent downward into the freezing heart of this machine. In they went, out they came. Numbers flashed across a digital readout screen. Each core was repeatedly analyzed, and then the long process of removing spurious magnetic overprints began. The cores were heated in steps, first 100 degrees Celsius, then 150, and on up in 50-degree increments. After each heating, which took about one hour, the core was again analyzed. The obscuring overprints of heat were stripped out of the cores; gradually an original signal could be seen, and if all went well, after many hours each core would provide one datum point. Because the Karoo rocks had been so subjected to heat over their long ages, they had to be heated to temperatures far higher than normally used. Hour after hour, the cores would be sent into the oven, cooked for an hour, and then measured in the magnetometer, step by step. The last step was near 800 degrees.

Our days were monotonous and not in the least romantic, the tedious downside of science that is usually never seen or publicized. We worked, went out to eat, worked more. Our great break each day was an hour's swim in the Caltech pool. Slowly the work progressed; slowly the many cores repetitively put into the oven, cooked for an hour, removed from the oven, run on the magnetometer, and then put back into the oven once more approached the temperature at which they would yield the information about the magnetic-reversal history of the Karoo and give us the best measure of time.

By now hundreds of hours of work and many thousands of dollars of precious research money had been dedicated to this project. We were sure that we were on the cusp of an important finding, whether or not the Permian-extinction boundary in Africa was of the same age as the one (as recognized in those rocks deposited in the sea) found in China, Iran, Europe, and Pakistan.

I flew back to Seattle before the long process was finished and readied for a new assault on the Karoo. I was going back to help Roger dig up the Gorgon. This trip would be financed by a television crew, and hence our ability to research would have to be tailored around the requirements of filming.

Several days before leaving, I received a phone call from a very distraught Joe Kirschvink.

"We had a fire," he moaned.

Okay, what sort of fire?

"In the lab."

I waited for the other shoe to drop. "And?"

Silence. Then Joe blurted out the news. "All the Karoo cores have been melted. Ruined."

I reeled.

Joe, ever a lover of innovation and new gadgetry, had been trying out new walkie-talkies in his lab, talking to his grad students in various parts of the building to test range or just to play—who knew? What Joe did not know is that the on/off and temperature gauge in his paleomag oven—coincidentally running *all* our Karoo cores at that time—was attuned to the same frequency as the walkie-talkie Joe was using. Unbeknownst to the humans in the lab, the tiny brain of the oven turned up its heat. By the time anyone noticed the huge amount of heat in that corner of the lab, the inside of the oven had reached twelve hundred degrees, hot enough to melt rock. All of the Karoo cores became liquid magma, sad little puddles of rock. Needless to say, all information was lost from them.

I listened to this tale and knew the bottom line.

We had to drill Lootsberg Pass a third time.

Damn.

■ ■ ■

The merging of the film crew and the paleontological crew was not a great success.

Another long trip to Africa, this one starting in March 1999, in the frigid prairie of Alberta where the film crew originated. We eventually arrived in Cape Town, where all the mean white customs agents who had been here in 1991 had been totally replaced by polite black ones. We spent a first hard, jet-lagged night in a Cape Town hotel, and then we made the long drive to the Karoo. Several days after starting this voyage, we arrived, in our many vehicles, at the Caledon River site where five months previously we had reburied our Gorgon.

On the last few miles approaching the site, we were all nervous. What if some rival collector, or just some random bone hunter, had found our cache and removed this priceless skeleton? But our arrival was anticlimactic; the cairn we'd arranged to look like a human burial site was seemingly untouched. The long dry summer had not disturbed a single rock as far as I could tell. Roger wasted no time. He immediately employed his crew in removing the rocks, and very quickly the skeleton lay exposed to the African sky once more, or at least as much of it as we had dug out on our last visit. But by now the sun was dropping low in the sky, and the short twilight would give us little time for any more work. We called a halt and headed back to Stone House for dinner.

The intervening time from November to March had been a busy one for the publicity people at the South African Museum. Gorgons are the prizes of South African paleontology, and up till this point a Gorgon with its complete skeleton intact was unknown to science. To have such an exquisitely preserved skeleton made the discovery all the sweeter, and great play had been made in the South African press about this find, even though it still sat in the rock.

Several artists' reconstructions of the animal appeared in newspapers and magazines around the world; various television crews lined up to record the recovery of this "new" (at least to the public) top predator. It was hard to persuade the media that this was not a dinosaur. *How could anything be older than the dinosaurs?* went the common complaint.

The press was invariably disappointed at the size of our beast. Something the size of a lion is not up to snuff when compared to a forty-foot-long dinosaur. But views of other Gorgon skulls usually converted the suspicious, for the ancient Gorgons were clearly fearsome predators in the purest sense of the word, perfect as B-movie horror icons. It does not take great imagination to get a sense of what it might be like having one of these monsters tracking your spoor, sizing you up for a meal, an impossibly reptilian lion with relatively short legs springing on you in lizardlike fashion. The last mad rush, then the slavering jaws with giant canines slashing through your bowels, spilling your vitals onto the rock, the final moment of life as the huge saber-toothed jaw closed on your windpipe—yes, the press wanted sensationalism. And Roger and I gave it to them.

It did not hurt in the least that this find had been made on American National Science Foundation funds, modest as they were at the time. We gave full credit to International Programs at NSF, and soon after, I was awarded a three-year grant to continue my work in South Africa. Now we were here on the banks of the brown Caledon River to bring this skeleton back to Cape Town. It would not be easy.

Our crew was so much larger than any we'd had before. Roger needed muscle to complete the weeklong job of excavating, plastering, and then loading the large skeleton onto a truck for its journey back to the South African Museum. He brought our crew from the previous trip, including Hedi, Georgina, and Paul, as well as another worker, Kerwin. I had brought the director of the film project interested in this Karoo work, plus a cinematographer, a sound man, a producer, and the acclaimed artist and my close friend, Alexis Rockman, from New York. Alexis was on hand to reconstruct the beast and paint it as it came out of the rock for the camera. This added contingent strained the accommodation ability of the South African crew—dinners had to be so much bigger, more food had to be bought. It was a change of the routine, and that was not to Roger's liking. We would all eat together, but the film crew would bunk ten miles away, in small cottages rented from the Tussen die Riviere game park. We expected to be there a week together.

Our first full morning of work began in curious fashion, at least for the Karoo: It was cloudy. We wheeled through the park, across the bridge to the Caledon, and pulled into the large field where the Gorgon lay. For an hour the film crew photographed the skeleton, interviewed Roger, and then packed up gear and asked to be taken into Bethel Canyon to see the Permian-extinction boundary beds, leaving Roger and his crew to begin the slow and painstaking job of excavating the great skeleton. We lurched across the rough track in two trucks.

As on our previous trip, the journey to the canyon from the Gorgon site took about a half hour of slow four-wheeling. The worst part of the trip was traversing a large dolerite ridge, rising about a hundred feet above the valley floor. The track across this was in terrible shape and involved great bouncing of the vehicles as they climbed over strewn rocks—not a voyage for the faint of stomach. It was a chastened field crew that arrived, finally, at the base of the broad canyon that Joe Kirschvink and I had so laboriously, and uselessly, drilled eighteen months before this current trip.

▪ ▪ ▪

Filming anything for documentary television is a mixed blessing. At first one is just so flattered at the attention. But very soon the reality of the many retakes and the nearly endless waiting—for one missing piece of gear, the correct lighting, or a sentence needing to be said over and over until it's just right—becomes apparent. It turns from fun to real work.

I was interviewed at the boundary beds. Filmed next to skeletons in place, I pointed out the changes from Permian to Triassic. In this way the day passed quickly. It was early afternoon when I noticed that the weather was rapidly deteriorating. A heavy humidity settled on us, hot but breathless, and dark clouds piled ever higher in a sky normally blue and clear. A cool wind came whistling through the canyon, and in the far distance a tenuous first rumbling lent low bass undertones to the day. Ominous ones.

I was no great judge of Karoo weather, but I'd seen everything here except a big storm, and I had no doubt that, like everything else, the Karoo was capable of violence on that score. I judged that such a storm was approaching. I yelled for the film crew to pack up. They gathered gear, stowed cameras, broke down sound booms and mikes as the first fat drops began to fall, and still it got darker. It was raining heavily as we finally got all the gear and all the people into the car. Our first thought was to wait it out, but I warned about getting out at all if there was a great rain, and got us rolling down the trail.

The world turned bright white, and the boom of the thunder was just behind. The storm had pounced into the wide valley, rolling over the high canyon walls as darkness and rain. Another white flash, an exploding crash, and we all became fearful. This was a fierce lightning storm, and we still had to crawl over the dolerite ridge. For a few minutes, we would be the highest thing in the valley—and metal to boot.

Thor was flinging his bolts with gusto now; great forked and anastomosing bars of electricity blasted downward and struck the Karoo hills around us, cleaving the nearly black sky with intentions of apocalypse. We crawled over the ridge, limped through the gates in blinding rain, and the track began to become a quagmire. Visibility was virtually nonexistent, the gloom punctuated sporadically with lightning bolts. For the first time in my life, I felt a fear of lightning. It had always been a rare joy in the past, but here it was a little too close to be impersonal. Two more gates to get through to reach Roger, assuming he and his crew were still there. Another smash of lightning, off in the field to our left, and I fancied I could smell ozone. Sure, why not? Of course the Karoo could do this. It had already been hot and cold and snowy—everything but this. Let's get it over with—*smash* to the right now, we were bracketed like soldiers in a rain of artillery.

We passed through the last gate, and just ahead was the Gorgon site. The crew there was scrambling around like ants in a flood. I jumped out of the car and ran to where Roger was feverishly working. A small stream was now running right over the bones, and mud was piling up around the skeleton. Roger was ordering his troop to build a small coffer dam up the hill from the site, to try to divert the water. The rain was pelting, and within minutes I was soaked to the skin, cold in the frigid wind, now carrying large rocks to the site of the diversion. I heard the loud crack of another bolt in the nearby field, and, while lugging a rock, I grunted to Roger, asking if it might not be a good idea to head out of here. Same everywhere else, he muttered, or something to that effect. So we worked, and soon enough the storm moved on, its bolts blasting other farms in its track.

We had done all we could do. The site was a mess. Where our cleaned skeleton lay but a few short hours ago there now rested a pond of red mud and debris. Roger had caught a Gorgon toe bone that the stream had begun to carry off at the start of the flood, and he worried that other delicate bones might have been carried away.

It was a disconsolate and bedraggled crew that arrived at Stone House for supper. We had no idea what shape the Gorgon would be in the following day. We cooked and ate, and the film crew headed off for their huts. I stayed at Stone House with the paleo crew, wondering what my mission should be on this trip. Lightning flashed in the distance, no doubt poking holes in Africa.

▪ ▪ ▪

The next morning broke clear. Paul was up early, listening to Radio Algoa, the single station that we could receive on the car radio. At breakfast he gave us the news that lightning had struck two people several miles from us in the storm of the day before, killing both. It could as easily have been us.

We returned to the Gorgon site and spent the day, literally, in mop-up. Oceans of red mud had inundated every crevice of the dig, and Roger lost a day in this cleanup. The film crew was happy; this was added drama for their story, but Roger was livid. How dare nature defy him? At first I was a bit amazed by his attitude, and then I realized that Roger was himself a force of nature. Time began to fly by, as it will when there's a deadline. Roger had his crew excavate, and they began to remove hundreds of pounds of strata and sediment from around the skeleton, digging an ever deeper trench around it. The work was brutal, for the Karoo's capricious weather had brought heat and humidity, and the sun became a stinging furnace. Kerwin suffered heatstroke by early afternoon, and we had to move him into shade. He was incapable of (or unwilling to) work after that, staying behind each day in Stone House.

On our third day, as the skeleton now lay exposed and ready for its first coat of plaster, Georgina wrenched something moving a heavy load and had to be evacuated to a doctor for help. She was in agony for days after with a badly injured lower back. Unable to sleep on the hard floor that served as our bed, she joined Kerwin on the sidelines, helping where she could, living in pain without complaint.

We all became sunburned, for the site was without shade, and the work demanded days in the sun without respite. Paul was our anchor, a huge man who could move a ton or more of rock a day. Wielding a pick or giant jackhammer, he worked twelve-hour days, his brown skin turning red and black, doing Roger's bidding, wresting this skeleton from the gripping rock. He quietly sang hymns. I became closer to Paul during this week, learning about his life in the evenings, about being born and raised a "Cape colored" in a country where color mattered more than in most. He had lost his beloved young son to cancer some years earlier and had found religion as solace. I had never seen such inner strength before, or such inner tranquillity in the face of unbelievable tragedy. He talked about the last days of his son, of the undertaker preparing the grave, of the community closing around him, of a death from cancer surely the result of the egregious pollution in the neighborhoods where the Cape coloreds had been situated by the whites amid the industrial squalor and chemical factories surrounding Cape Town. How do you comfort a ten-year-old boy with inoperable cancer where there is no health insurance or adequate medical coverage for someone of your skin color? How do you avoid exploding with anger? If the boy had been white, he might still be alive. How do you not go out and blow things up in response? And that's exactly what was happening in Cape Town, in a bombing campaign against white targets.

Roger was never happy now, for the casualties were causing the work to fall behind schedule. The elements seemed to conspire against him. The heat never abated, and even the nights became a torture. Thanks to the rain, the mosquitoes had rapidly laid their eggs, and within several days we were beleaguered by them at night. One of the Canadians had brought a mosquito net, but the rest of us were sorely taxed each night by the perpetual whining around our faces. I worried about malaria. Sleep was hit or miss. On top of it all was the film crew, wanting this shot or that interview. It was a different Stone House from the place where we'd lived some months earlier.

It was Paul who came to me on the fourth day or so to tell me, sotto voce, to keep my eyes open. He was the only one who had any contact with the outside world, through his morning radio news show. A band of marauders was nearby, and a number of cars on the road we used daily had been hijacked. Several white farmers and their families had just been murdered in our area, perhaps by the same group, perhaps by another. I did not pass this information on to the film crew, for they were already as alarmed as could be. Nevertheless, each day when we passed a group of blacks walking by us on our road, I would wonder, and worry, and equate skin color with violence, and fall into the trap that had ensnared all of South Africa.

The film crew had had its share of misadventures already; the producer and Alexis, on a mission to get food in Bethulie, had gotten lost on the way back, and night had fallen. They had no idea how to find us and tried looking for help. The place they ended up in was the maximum-security prison in the area. They drove up, innocent as can be, and were astounded at the armed reception they received. Was this a jailbreak? wondered the guards. All was sorted out eventually. The guards gave them directions, and they found their way to Stone House and safety.

Roger was obviously under intense pressure. But as in all things that I could see when observing him, he persevered, and through sheer force of will, he succeeded. What is a minor detail such as a giant rock-encased skeleton that had to be removed from the Karoo to someone who, in his late forties, runs competitive marathons? The large skeleton was now completely plastered, and the entire block of plaster and skeleton weighed somewhere in the neighborhood of two tons. He had to figure out some way to get this great block up onto a truck. He eventually rented a crane, and the plastered skeleton was slowly raised. Just as it was being lifted out, one of the slings shifted, and the two tons of rock and plaster quickly fell over toward the side Roger was on. He rolled out of the way like a movie stunt man, narrowly avoided a grisly death, berated the lift operator, and started over. Up it went this time, onto the truck, finished.

A huge white object sat on the great flatbed truck. An enormous hole gaped in the ground. Like good golfers, we replaced the divot. The Gorgon headed for Cape Town, the film crew for Canada. I went to Lootsberg for another look at the boundary beds, before going back to Cape Town, for one free day before heading home.

▪ ▪ ▪

Several weeks after getting home, I received an e-mail from Roger. His preparators had taken the plaster off, begun on the head region, and had a shock. The Gorgon we'd worked so hard on wasn't a Gorgon after all. It was something entirely different—a bizarre herbivore possibly related to *Lystrosaurus,* perhaps the ancestor of the *Lystrosaurus,* in fact. In any event it had become a false Gorgon.

I thought that this might be a very trying discovery to Roger—for the press already knew this specimen well. But Roger shrugged it off. After all, there have been many Gorgons found—he and I had found a splendid head at Old Lootsberg, he pointed out. But this giant skeleton, this false Gorgon, was from well down in the Permian, something that no one had ever seen before. (Eventually it would be completely removed from its rock, becoming one of the prize specimens of the South African Museum.) Yet it was somehow connected to the archetypal Triassic animal, *Lystrosaurus.* It was telling us something of significance.

Chapter 9

The Rate of Killing

OCTOBER 1999

The years had been passing by; seasons flew, and we all grew older. My father died in this interval, having spent his last week in a semicomatose state after suffering a stroke. I sat at his side watching a tube pump bile out of his system, watching him die and waiting for a final benediction from him that this journey I was on was worth its effort, worthy of his respect. That never came. I watched him, saw the terror in his eyes, saw him leave his body, wondered why I did what I did, and wondered what the sum of his life had been, in those last moments of life as he knew it. Could simple curiosity be all, or is it fear that we all go away with?

And then another death of someone close to me, as shocking (if that's possible) as the death of my father. For a decade I had been watching the progress of someone truly called to paleontology, a prodigy of a schoolboy from a nearby town who began his career collecting fossil leaves and then moved on to other fossils. When he was ten, I began to take him on my class field trips; as a young teenager, he continued to blossom and corresponded with all and sundry. He published his first scientific paper before he was fifteen. His parents moved him swiftly through school, and I met him again in my senior-level geobiology course in the spring of 1999; he received the highest grade in the class. He had passed up high school to jump directly to college, and at seventeen was already of junior standing in the university. He was tall and robust, had a great sense of humor and the emotional makeup of an adult as far as the world could see. He was going to change our science, and we professionals now greeted him with respect and welcome.

The first hint of trouble came that summer, when I invited him to come with me to the Karoo as my field assistant on my next trip. He demurred and said something about possibly joining the air force. This was shocking news; I asked him to reconsider, but knowing how he had been hurried through life to this point, I wondered if that might be a good thing for him. I did not know that he was harboring a secret, to him a shameful secret: He had done poorly in freshman calculus and thought, incorrectly, that this would close the door to the field he so wanted to join. I could have given him a good laugh on that score had I been given the chance; my scores in calculus were absolutely abysmal, yet I prospered, as did many others in my field. But he never came to me. He never went to anyone. He never learned that we all fail, in greater or lesser ways. He had a shining future but, to him, a bleak present. Over the summer I sent him the occasional e-mail, but I knew he was traveling with his family. In September I received an invitation to his funeral. He had shot himself in a fit of despair. I was named in the suicide note; he did not want to disappoint those of us who had been beacons in his life. I still miss him.

His picture is hung in my lab, and new generations of students have already forgotten him. They see only the portrait of a husky youth, smiling, a leader, and wonder why he is on that board, with Jack Sepkoski and Steve Gould and the other dead paleontologists who changed the way we look at the past.

■ ■ ■

Why does time when lived pass so slowly, yet time remembered exists as a blur? For the Karoo research, our progress was measured in years, in fits and starts, in small victories and seemingly large failures; so, too, with life. By 1999 we knew a great deal that was unknown in 1991. But had I known how hard the journey would be in 1991 would I even have embarked on the road? The resolution of the Cretaceous/Tertiary extinctions as being caused by a meteor was relatively quick compared to this current work. Was it the greater age of the rocks that caused progress to be so fitful? Fewer places to study? Fewer investigators making the effort? A cause that was less amenable to discovery by scientific methodology? All of the above? It now seemed a project that just dragged on, with no real end in sight, in spite of the victories. The possibility that we would never really know what caused this greatest of mass extinctions, what brought the Age of Protomammals to such a swift and dramatic end, was becoming increasingly real.

There were two solutions: give up or keep going. It was a nobrainer. We had money now, real money. The National Science Foundation had given Joe Kirschvink and me enough funds to take three more trips, enough to defray the cost of both the South African Museum teams under Roger Smith as well as the Bernard Price Institute expeditions of Bruce Rubidge and John Hancox. So we kept going.

My trip in 1999 was again in the autumn of the Northern Hemisphere, the early spring of the Southern. I left Seattle in a terrible state of mind, frantic with the last-minute minutiae of leaving for a long trip, grieving the death of the young boy, playing the "what if" game. My last day before this long trip was filled with finishing unimportant things and not getting to the things that mattered, such as a good-bye to my two-year-old, good-bye to my family. I didn't want to think about the real perils of traveling to South Africa and dealing with its violence and crime and road carnage, for the death rate from automobile accidents is the highest in the world. My wife dealt with the stress by taking trips to the dump and painting doors and dancing around the real problem—the impending distance. Was science really so important that I needed to go away for a month, leaving her alone with a full-time job and an infant to care for? What the hell was the use of finding out what had caused some long-ago extinction anyway? Who would really care? And I would tell her that it did matter. Deep in my heart I knew it mattered, but I didn't know why I felt that way, and I wasn't able to express these feelings with any success. So hard words were said as I packed, and my small son watched the hateful suitcase fill with field gear and cried at my leaving, clutching my leg: "Daddy don't go." I shook him off as gently as I could, but there is no gentleness in this sort of leaving, only scars, scars that come back to haunt us in the dark of night when sleep mocks and we take the measure of our lives. Why do we do what we do?

I drove to the airport angry and empty and in pain, emotional as well as physical, the latter from a pinched nerve in my arm from a basketball game the week before. I boarded a completely packed 747 bound for London and got a middle seat next to an extremely overweight man who managed to ooze into half my seat (and onto me) as well as overwhelming all of his own for the ten-hour, all-night flight. I snarled at him to stay off me, and he physically could not. I arrived in London at 11:00 A.M. London time, and, having to wait for a 10:00 P.M. connection to Cape Town, I took the tube into London just to try to stay awake. I walked like a zombie through art exhibits and finally came to my favorite painting by Seurat at the National Gallery, the pointillist bathers all composed of small dots that up close mean nothing and from a distance distill into a faraway place. Moving on, I fell asleep in a movie theater, then ate some bad London food. London was chilled, and I made the very smart move of buying a pair a gloves on the street as a cold October sun set behind Piccadilly. I didn't know it at the time but the gloves would come in handy later in the African cold, and they felt good on my fingers in the gathering London twilight. Finally I boarded my flight to Africa in the dark London night, again a filled 747, again in a middle seat, this time with boisterous African seatmates. The unending flight was again sleepless as we flew from the far north to the far south, crossing the equator in the dead of night. There is romance to science, but flying isn't part of it; scientists fly tourist class, and I can't sleep on planes.

I arrived in Johannesburg at eleven in the morning, twelve real hours after leaving London, and now around thirty-six hours after leaving Seattle. I had two hours to kill in the J'burg airport, trying to stay awake to guard my money and valuables before I boarded a one-hour flight to Port Elizabeth, at the opposite end of the country. I grabbed my heavy bags, found the rental-car agency, panicked over the route out of the city, briefly got lost, did a U-turn in major traffic, located my road, and at three in the afternoon headed away from the tranquil swells and beaches of the Indian Ocean and out of the Port Elizabeth, driving on the left side of the road past the squatter camps and the seas of garbage surrounding the city, toward the interior of Africa.

For the first hour, the land was green, fertile farms and forests in the mountains near the ocean, but eventually I left the hills, and the countryside became increasingly brown, and then I was in the Karoo. Dusk was approaching as I rolled into the sleepy town of Craddock, bought fruit and juice and caffeine, and was surprised at the overwhelming cold. Dark clouds had covered the sky, a rarity in the Karoo, and a cold front had moved in from the southern ocean, perhaps from not-so-distant Antarctica. The chill was visceral and unexpected. I drove on and in my third hour began to see familiar signposts, names of towns I had passed on previous trips, all the stranger for my days of travel and the darkness of the night, places that I was used to in the light of day.

I finally began a long descent into the giant Lootsberg Valley, now in total darkness. Roger had told me to meet him at Tweefontein just at the base of Lootsberg. I left the highway, had to navigate the back roads in the dark, and entered the land of farms where each section is closed off by a wire gate. I had to jump out of the car at each, open the gate, jump back into the car and drive through, jump back out of the car to close the gate, back into the car, and then back on the dusty track, rolling my rental car across perpetually rutted farm tracks. In the absolute blackness of this African night, gate after gate after gate, I hoped that I was on the right road to my friends, to food, to a camp, to warmth. I was imagining luminous eyes at the sides of the road, and snakes, and I wondered what was real anymore, and what I would do if I could not find the camp. Roger had my camping gear with him.

I passed one more gate and then saw a small twinkling fire in the distance. It was now about seven-thirty, and I rolled up and saw Roger bending over a grate on a fire in front of an old farm building, casting bizarre shadows as the fire shuddered in the strong wind. I got out of the car into a hard cold wind, my legs cramped from the long drive, and saw Georgie, Hedi, and Paul coming toward me. I knew I was home. I hugged each of them, taking their measure and they mine, noting my fatigue, genuinely happy to see me. I gave Paul an extra hug, for I had missed him so, and smiled my happy smile at Georgie, once again under the African skies amid memories, with new ones to forge. Roger was finishing a delicious dinner, and this new home was a large camp tent with a table and camp chairs, iceboxes, and an area on the ground for sleeping. I so needed a shower, but the only shower was a hose hung over the side of the decaying farmhouse that was connected to the irrigation system of the nearby fields. Ice was forming on the hose as I went to look at this contraption, and I decided to skip the shower. Once our meal was consumed, I spread my bag on the floor of the tent and crawled in, all my clothes on against the cold, sleeping with my mittens in an REI sleeping bag supposedly designed for low temperatures. And so I arrived. I had changed worlds again, from the world of Seattle to the world of travel to this world of the Karoo.

By morning I felt the relief of a whole night's sleep, none of the waking-at-three-in-the-morning nonsense that usually plagues me on long west-to-east plane trips. I was awakened by Roger, bearing a warm cup of tea; I looked over to see the picture-perfect face of Georgina in the sleeping bag next to me; Paul was coming in carrying his coffee and gently bidding us a good morning. Delicious smells of breakfast: bacon, eggs, toast, juice. I began to emerge from my sleeping bag and was shocked by the cold. A howling wind was blowing outside the large tent where I'd slept, and where we would take our meals.

I had to have a shower. I walked out behind the old farm, stripped off my clothes in the howling wind, grabbed some soap, and stuck my foot in. It was brutal—a shock of freezing water from that rigged-up hose, cold soap not foaming in the hard water, shampoo half frozen on my head. More water on the legs, then the chest, take a breath, over the back, final breath, do the head again for a rinse. Shock and pain. My body jerked awake and covered with goose bumps. This was what death must be like. Rough drying with the small towel, my body white and ghastly looking in the cold. I rushed back into my clothes and began to ready for the day, shaking.

The first day was the hardest—the jet lag, finding that first fossil, leaving one life behind and beginning another, even finding the right gear. I needed so many individual items of gear for the Karoo research: sunglasses, hat, Brunton compass, GPS, Swiss army knife, watch, water, backpack, hammer and hammer holder, eye loupe, glue, sample bags, lunch, fruit, suntan lotion, lip balm, field notebook, sample number slips, pencils, pens, felt-tipped markers, boots, double socks, blister skins. All would be necessary sooner or later in the day, assuming I found anything. All had to be collected on this first morning and put in correct order.

The activity in camp increased; lunches were made as breakfast dishes were washed. We began to mill around the truck, and Roger marched out, jumped in, and fired her up—a signal for us to clamber in. We were off to the field. It was 8:00 A.M.

Our field area was several miles to the east of camp. As he drove, Roger gave me my job. He had seen some interesting low bluffs in the far distance and wanted me to examine them for fossil content. We lumbered over the farm road I'd taken the night before, lurched through the several gates, and now took an even smaller track, for "road" was not a word that befitted this path. The day was splendid but cold, like a winter morning, and after about fifteen minutes of slow lurching, Roger came to a stop. I was being dropped off. He gave me the rendezvous spot, some miles distant on the next farm. I squeezed out of the cramped truck, bade my companions good-bye, and tried to gather my gear around me. The door slammed, Roger threw the truck in gear, and it lumbered off in four-wheel low. Within minutes I was alone on the vast grassland.

Several emotions hit at once. First, the absolute loneliness of the place. I was in a gigantic valley, at least ten miles across, in air so clear that even the distant sides of the valley walls were visible. In front of me the great mountain range making up Lootsberg Pass dwarfed all. The incessant wind caused waves of motion in the taller grassy tussocks. I remembered another time of being dropped off like this, long ago in New Caledonia, when a friend of mine and I were dumped out of a small boat in a wind-tossed sea for a scuba dive. The sea was too rough for our boat driver, who had piloted a small, motor-driven rowboat at least five miles to let us off. He told us that he could not stay out here but that he would be back in an hour to find us. We were summarily left in the water. Good as his word, he headed back toward land. There, too, I was in a great empty sea, with only the gear necessary to survive and the need to start. I now began to walk.

The outcrops were low and spotty. The name of the game was to spend as much time as possible on rock. There were no fossils in the grass, among the bushes, in the tussocks; it was only bare sedimentary rock that mattered. And when you are actually on rock, your head as well as your body must be there. On this first part of my journey, there was very little rock, and I walked faster across the rough terrain to get to the low blue outcrops I saw in the distance.

In this way the morning passed. Near ten I found my first scrap of bone, after more than an hour of looking. It was a partial skull, but the rock it set in had been heated by a nearby dolerite dike. Was it identifiable? I marked the place and then continued on.

In my second hour of searching, I found a small string of vertebrae and feverishly searched for a skull. But this body had no skull with it; it was headless, and thus would be forever nameless, for the head gives the identity here.

Hours had now passed, and I was still far from the rendezvous point. It seemed like another world, another way of life to be out in the middle of this giant sky, no other humans around, only my two legs to carry me to my friends. I had walked perhaps two miles now, but in the slow searching walk of a paleontologist, not my normal stride, and I reckoned that the truck and the rest of the crew were over the next hill. I climbed up and into a separate small valley. With my binoculars I could see them, perhaps a mile away. By this time I had worked and searched for better than half a day. My efforts had been fruitless as far as identifiable material was concerned. It was late afternoon when I rejoined my fellow collectors. Roger had found two skeletons, and Hedi had found the head of a huge *Lystrosaurus*. All of these data were important for defining the boundary.

I was dying for sleep. I barely made it through dinner, and by 9:00 P.M. I was out, as the camp bustled around me, people washing dishes and preparing for the next day. The jet lag on the second night was excruciating. I woke up at 3:00 A.M., my night over, but could not switch on a light to read because of my companions sleeping around me. The wind was rustling through the tent, flapping canvas doors, mocking my attempts at more sleep. For hours I lay there, thinking, not sleeping, waiting for the first call of birds, the first hint of dawn, the end of this night.

Of course I finally fell asleep again when dawn came. I was roused out of my sleeping bag, stiff from sleeping on the ground, pad or no pad (it felt like no pad), grumpy, fifty. The others laughed at me; they took me as seriously as I deserved. *You are such an important man! You think you can sleep through breakfast! Out of bed! Lazy American! Good-for-nothing!* Delicious smells filled the tent. Roger had combined eggs and leftover meat into something heavenly. I had granola as well; South Africa makes the world's best. It was all fuel for the day to come.

We had no outhouse at this camp. A shovel stood in the corner with a roll of toilet paper wound around its handle. Everyone found some remote spot behind some tree or other. It was awkward business on those cold mornings. We pretend we are not animals, with our antiseptic and comfortable bathrooms. The Karoo lays bare the lie. I mourned our Stone House. I mourned any house, still feeling sorry for myself in this cold.

We loaded into the truck and were off.

▪ ▪ ▪

The second day we worked in the region of Wapatsberg, a high pass several miles from Lootsberg and several miles from camp. I was here the year before with Joe Kirschvink, drilling. We arrived at our field site at about nine in the morning on a day that had dawned very cold. Over breakfast Roger looked at the gray sky and, optimist that he is, predicted snow. We dressed as warmly as possible, covering ourselves in various layers: long underwear beneath jeans, a T-shirt beneath the heaviest sweater I'd brought, the heaviest jacket, gloves, a wool hat. This seemed such an odd getup for Africa, for the desert, for a place so famous for its heat.

We drove up the old road past the farms, through the various gates, and finally up high in the hills, with the most breathtaking views. The sun had broken out behind a great bank of clouds, but a strong wind still blew fiercely, and the cold had a grip on the land. Roger gave us a pep talk as usual. He told us what the bone might look like; he told us how the bone in this part of the highest Permian is generally found; he told us that there would be few complete skeletons, perhaps only isolated skulls, and the rare, large, postcranials, too. We jumped from the truck, staked out territories in the vast panorama below, and started off. From the high road, we split apart, climbing over the barbed-wire fence and spreading out, isolated now, each of us heading for a patch of outcrop, a region to begin research.

I spent an hour searching in outcrop just below the road. The outcrop was perfect for finding fossils. Yet even with this beautiful exposure and even on this early morning with my mind clear, there was absolutely nothing to be seen. We were working in the region that overlaps between the *Lystrosaurus* zone and *Dicynodon*. It is the most critical region to answer the most critical questions: How rapid was the mass extinction? Did it occur at high standing diversity? Did all the taxa go extinct simultaneously, or was it a long, gradual diminishment of diversity—did the mammal-like reptiles, one by one, go extinct over hundreds of thousands, perhaps even millions of years?

It was not until I'd been searching for nearly three hours that I found the first scrap of bone and then the fossil gods, as capricious as they are, yielded a small skull in place. What jubilation! After long periods of time finding nothing, it is so easy to let the mind wander, to think of other things, to enjoy the gigantic panorama of blue sky, the white clouds whispering past, to think of ways of banishing the cold. So easy to think of home. I felt a huge and painful swell of regret, guilt, homesickness, longing for my family, and a need to end these pursuits. And then, having felt sorry for myself for a while, I put it somewhere else, kept looking, and found this skull.

With the first find of the day the heart quickens, the pulse races, excitement is visceral. There is also a new problem—extraction. Can I remove this fossil, or should I wait for Roger? Is it worth his time? Is it a small piece, part of a larger skeleton, as I first suspected, or is it one of the many red herrings, the numerous disappointments, the great discoveries that turn out to be not more than a scrap of unidentifiable bone, or nothing at all. I chisel around the fossil carefully. It turns out to be a very solid, small skull of an unknown dicynodont, and I mark it as Roger has ordered us to: a nearby rock with a piece a toilet paper underneath it. There is so little rain in the Karoo that these toilet-paper markers can last entire seasons. In the past we wandered through areas and found these bedraggled markers, and we knew for sure that some member of our crew had once found a fossil here.

The skull was now marked, and I felt the joy of having one in the bag. I would not go home tonight with nothing to show for a whole day's labor. It was a small victory, only a single mark on a growing list of recovered fossils, but a mark nevertheless. A mark that I'd made, and I was quite content with this small find.

Nearly half the day had now gone by. It was time for lunch, and the search for a comfortable place to sit down commenced. I wanted to remove the pack and get off my feet, but the ground was composed of nothing but sharp rock and thorny brush and, covering all, an infinite number of ants. Small ant nests were everywhere, and I knew from sad experience that a lunch site must be chosen carefully, away from any of these ferocious armies that took such umbrage at large humans' sitting near their nests.

Lunch is always a treat in the bush. On good days it consists of leftovers from the night before, because crew suppers are always exquisite things. Otherwise a generic sandwich and a piece of fruit. If I'm lucky, I've remembered some sugar of some sort. Long draws of water, a chance to rest tired muscles. Respite. Brief respite.

Ten minutes went by. The lunch was inhaled. And then I was back to it. It was time to go.

The fickle weather began to change again. Up until now most of the day had been cool, with great fleecy white clouds covering the sun whenever it threatened to warm up. Soon after lunch the cloud cover began to dissipate, and on the bare rock the day warmed rapidly. The coat, sweater, long underwear—all had to be ditched. Shedding clothes was a chore. I hated to take my boots off in this thorny, rock-strewn land, but my long underwear, was now roasting me. I changed into the shorts and T-shirt I'd been carrying in my backpack and resumed searching for fossils, much invigorated by the lunch, the find of the morning, the brief rest, and the clear warm air of the Karoo. The sun was a tonic after the cold and gloom; it raised my spirits, and elevated my mood. Wild mint lined the creek beds, and the air became redolent with its sweet smell. The entire valley was laid out before me in its perfect desert detail. Life seemed ideal.

I searched along the hillside at the base of Wapatsberg Pass, the stacked strata and overlying dolerites rising a thousand feet or more over my head. The rocks are alternating mudstones and sandstone, and I concentrated on the greenish to gray mudstones, trying to imagine them back into a 250-million-year-old-river valley with swamps and shallow lakes and copses of green trees lining the banks. I imagined a quiet place, rich in vegetation, and rich with herds of dicynodonts. But such musings are the enemy of fossil recovery. When the imagination wanders, so do the eyes. You can walk right over the most beautiful skeleton, yet if you're in deep thought, nothing will be seen. And so in these early-afternoon hours, I had to search with ever more vigilant concentration.

Another hour passed, and then, near two in the afternoon, I found another fossil, the jaw of a small dicynodont in place, again a cause for celebration. I marked this fossil with a small wisp of toilet paper, noted its position with my GPS, and set out again, now fully pleased with myself. My reward was a pull from the water bottle, for I had no food left. Then I noticed with some alarm that the warmth of the last hours had once again changed, as piles of tall gray clouds blew up from the south and started covering the high pass at Wapatsberg, plunging all into shadow below. The wind blew stronger, and a fine mist began to fall, growing to a steady rain. My thin T-shirt quickly soaked through. Fool that I am, assuming good weather for the remainder of the day, I'd left my warmer clothes beneath the tree where I recovered the first fossil, and I was now more than a kilometer away from that spot. I set off marching across the rugged rocky countryside as fast as was safe, hoping that I could quickly find the ravine where I'd left my clothes. Such was the nature of the small victories and small defeats of a day in the Karoo.

The last hour and half of the day were uncomfortable and unproductive. Although we searched in the cold rain, the wet, rocky surfaces gave no glimpse of fossils. Across the vast swath of countryside we worked, I could see through my binoculars my compatriots scattered at great distances, small twig figures engaged, like me, in a wet and ultimately futile search. We gathered at the car at five, another day finished, and drove back to the camp a few fossils richer for our hours of search.

There was quiet satisfaction back at camp. Roger knew, and I knew, that we were succeeding. Slowly but surely we were wresting a secret from the earth, from time. The mammal-like reptiles died out quickly, not gradually. Day by day the numbers confirmed this. We were on the home stretch now.

■ ■ ■

We'd begun card games after dinner as something to do. We played in our large tent, at the meal table. The stakes were imaginary and preposterous. On one such night, for reasons unknown, the tent was invaded by gigantic flying cockroaches. They were monstrous creatures, out of a bad sci-fi movie. Hedi refused to kill any bug, so she carefully tracked them down and released them outside the tent. Our bright propane light was just too powerful an attractant, however, and within minutes new ones were back, along with a staggering diversity of smaller insects. As we turned out lights, a number of the monsters were found in our sleeping bags. Pandemonium, laughter, lights on, bugs out. With the lights on, I looked at the floor of our tent: two ticks on a mission. Hedi or no Hedi, they were executed.

Throughout this trip I'd been drinking only bottled water. My colleagues humored me. They were drinking from a nearby well.

But after a heavy rain, the well was filled with runoff from the sheep and cow fields, and Roger was hit with an excruciating intestinal ailment, no doubt from this bad water. Seeing Roger sick was a shock to everyone—most of all Roger. The man is carved from rock, seemingly not a biological being; how could he get sick? There was no time in his schedule for this. Besides, he'd been making a point about how South Africans can drink the local water with impunity. But sick he was. He had a high fever, and finally resorted to some sort of nineteenth-century African or colonial Brit purge. I thought a witch doctor would do as well, but I kept my thoughts to myself. As far as I could perceive, Roger has never been wrong about anything in his life (in his view, at least). He moved himself to a separate tent for one night. He looked ghastly. He slowly recovered but was still weak. Nevertheless, he headed out into the field for work as usual. Everyone switched to bottled water, and ribbing of the weak-stomached American subsided.

On our seventh day in the field, I declared a holiday. Work was not stopped, but that evening after fieldwork we traveled to the nearest civilization, a lonely motel called Good's Bottle Inn. The entire front of the motel area was made up of glass bottles, and the prized possession of the motel was its bar, a fact that gave rise to ever more bottle fences. It was the local watering hole for a huge region, and the Karoo farmers (and the few highway voyagers who stopped here) drank prodigiously. There was definitely no shortage of bottles.

We were there for a shower. The proprietors were right out of a creepy movie, but happy enough to oblige our wishes for a nominal fee. We took turns and gloried in our first hot water in over a week. After showering we drove to Graaff-Reinet, forty kilometers away, and I treated the crew to a long, wonderful sit-down dinner in the Drostdy Hotel.

At the restaurant things began in festive enough fashion,

enjoying a bit of civilization. The aged barman brought out wonderful cocktails, and I indulged in my favorite Karoo drink, lime juice and tonic water. Paul, who did not drink any alcohol, was talked into having one of these Shirley Temple clones, and I was gratified by the big grin that spread across his face. But then it all turned bad. Roger, the only real drinker in the crew, went through much of a first bottle of red wine, then much of a second. He got nasty, which was very rare for him, and began to berate the crew for not working hard enough, not finding enough. We all knew that none of us were anywhere near his equal at finding Karoo fossils, but everyone had been working twelve-hour days in cold and heat without complaint and with unceasing, if slow, results. I listened, accepted a rebuke, but as it continued, I finally had enough. Through all of these years with Roger, I'd voluntarily subjugated myself to his authority and expertise. I'd been a silent partner, but an intellectual equal at least. I'd taken pains never to question his authority or knowledge in front of his crew; among ourselves, however, we fought scientific points with passion, and these fights were good ones, and necessary. But this type of chastising of a field crew would not do, and I told him so, in no uncertain terms, at the table. The crew looked on in embarrassed and surprised silence. They had only ever known affable Peter; they'd not before seen the carefully hidden wells of anger that were too easily unleashed, or unleash themselves. This night was for celebration and reward, not for bullying, and alcohol-fueled bullying at that. I would not have it.

The ride home was frosty, with no one speaking. Back at the tent Roger privately apologized to me—not about his comments concerning the inadequacy of his crew but for being so loud about it. I saw red again. He did not get it. These people would walk to hell for him—as I would, and had, on this project. We could do no better. How could he not see this? Or did he see this?

This man did not merely run marathons, he ran ultrama-rathons—after the twenty-six miles were up, he and a few other souls ran seven more. Was this what this project was all about—run the race, get to the finish line, and then run some more when the race was finished? Fuck that.

▪ ▪ ▪

The moon had been waxing these last few nights, and on this night it was full. It rose above the great rampart of Lootsberg Pass after dark, and I could feel its call even at dinner in our giant tent. After dinner, dishes washed, I decided to go for a walk.

I grabbed a jacket, and Roger, wondering what I was doing and hearing my romantic answer, gave some chary advice. "Don't walk on a puff adder," he intoned.

"Right," I responded. It was night. What would a self-respecting puff adder be doing out in this cold? Roger replied that they rest in the dirt roads on cold nights, the dust being better insulation against the cold than bare rocks, and will still strike if they are disturbed.

Sure. Bogeymen don't scare me.

All bundled up, out I went. Sure enough, as I emerged from the tent, the giant moon struck me with all the light it had redirected from the sun. It was amazingly bright in this clear, high air. Lootsberg is over a mile high, and at thirty degrees south latitude, I had a real show on my hands. But as I looked at the moon, I detected something very wrong—it was upside down. I experienced a moment of serious disorientation before realizing the obvious explanation—I was in the Southern Hemisphere. The old man in the moon was standing on his head. Such things are a reminder of distance from that which is familiar, that which is home.

I headed away from the farmhouse, and as my eyes became more accustomed to the night, the bright moon made it seem like daytime—dim daytime, but certainly not a normal night. The veld around me was multitextured, and the cool air begged a fast trot. I ambled down the track at a faster pace now, enjoying the feel of my body in this glorious night, and I was perhaps a mile away when Roger's admonitions about puff adders was remembered. Puff adders kill more people in South Africa than any other snake because of their nasty temper and lack of warning. They are aggressive, not defensive and retiring like the Cape cobras, which strike only as a last resort. Puff adders seek out trouble. Apparently so do I.

All of a sudden my preoccupation with puff adders caused me to lose sight of why I was out here at all—to enjoy a fabulous African full moon on a glorious night. So I rather grumpily decided to jog back to camp. But now every shadow hid a snake, every piece of dead wood on the road (and there are many) was a live serpent. My imagination was in overdrive. This was now misery. Damn Roger's eyes.

The next morning, driving out, there was a fat black snake just outside the first gate, a pustule of dark and evil coil. Puff adder. The most repulsive and dangerous snake on earth, set against a parchment of beauty. Beauty and the Beast. A perfect symbol of the Karoo.

• • •

In all of our time camping at Tweefontein, the old farmhouse remained a mystery, and its graveyard even more of one. I found myself gravitating there at sunset each night. As the sun dropped, the shadow of the low hills to the west of us would stride across the broad valley, and for a few minutes the black shadows of the three headstones walked up the side of Lootsberg in tandem, passing the Permian-extinction boundary like wraiths. They grew, stretched, and then were engulfed by a greater darkness.

On our last afternoon together, fresh from the field, Hedi came out and joined me. She was such a contradiction—by far the best fossil finder after Roger, a woman of undeniable and broad knowledge and talent. Yet she dismissed all praise and did not believe in the extent of her gifts, or seemingly even in their existence. I always felt comfortable in her company and asked her one evening to translate the headstones. We walked together to the iron-barred site, covered with weeds. The place had an aching air of melancholy. The largest headstone marked the grave of the husband and wife, Anna Catherine and Francouis Petrus. The inscription was in Afrikaans, of course, but Hedi, a native of Germany, knew enough Dutch to make out the inscription. It read:

Take free my dust of earth till dust covered again, one day I will rise again through God's almighty awakening.

The headstone of their son Carel, dead in 1898 was next:

Man's tongue glorifies God.

The last member of the family in the graveyard was son Jacobus, whose 1900 inscription stated:

God is the love of angels.

The sun was dropping behind the hills now, and a great blackness rushed into the valley, but the western sky filled with red and orange, the violent light giving the last visions of this family's grave, and the far older grave it coincidentally stands on. We turned back to camp, and I looked at the inscription above the resting place of Jacobus Fouche once more. It rolled in the mind, a sweetness.

God is the love of angels.

When that phrase was written on the grave of the newly dead youngest Fouche son, South Africa was in the midst of its Boer War, and the most unspeakable atrocities by both sides were being conducted around Lootsberg and elsewhere in the Karoo. What did the angels see in South Africa amid this slaughter, where tens of thousands of Boer women and children were rounded up, removed from their farms, and then allowed to die hideous deaths from starvation and disease in crowded concentration camps? Where Boers shot the surrendering British troops? Where the victorious government, at the end, stole all land from any native African and gave it to the white survivors? Is that what killed off this Fouche, a Boer true: Was Jacobus taken to the land of the extinct by a British bullet, or did he die a rotting death in a British prison camp? That scenario is the most likely one of all.

Chapter 10

Drawing Conclusions

Roger Smith arrived in Seattle in early June 2000 to finish our manuscript on the stratigraphic ranges of protomammals across the Permian-extinction boundary, the summation of the many years of Karoo work up to that time. He e-mailed me that he would arrive on June 3. He did not tell me where he was coming from, or the flight number, or even the airline company he was flying with. I sat waiting in my office for a phone call. It was Saturday. He never called. Sunday came, and I e-mailed his wife, now quite alarmed at his absence. She did not get the message. A call from Roger came finally on Monday. I retrieved him from an absolute dive of a motel near the airport—all that his meager rands could buy him for accommodation. He had forgotten to bring my phone number and had to wait until Monday to track me down. Somehow this screwup seemed fitting. Communication had never been our strong suit.

Roger brought the data from the many field trips we'd shared, the long years we'd logged in pursuit of the culprit that had caused an ancient mass extinction. Finally together, we sat in my Seattle office with the assembled data and slowly put together a range diagram based on our results. First we drew the stratigraphic sections from our notes, putting each distinctive rock type onto the diagram and placing all into a context of stratal thickness. Then we began the laborious job of transferring each fossil we had collected onto this stratigraphic context, so that each discovery was given a distance either above or below the Permian/Triassic boundary. One by one the tick mark denoting each fossil found its place on this chart. Gradually each species we had recovered was given a range in time. It seemed, at an emotional level, not in the least bit fair: each fossil, often the product of whole day's work, trivialized into a single, tiny mark on a chart. There should have been some larger message, a billboard celebrating each hard-earned find. But the tick marks kept sprouting, and slowly a picture emerged that lent dignity and closure to our work, or so it seemed to me: The measure of our labor was a message of rapid catastrophe.

It was a very curious experience to have Roger with me in America, in my hometown, staying at my house. He seemed smaller here, out of his element, subdued. Just as the Karoo seemed to enlarge him, so did my American city seem to make him somewhat frailer, another human being, not the force of nature I had become used to. He seemed genuinely moved that I let him stay in my home instead of relegating him to a hotel, always my fate when visiting Cape Town.

To this day Roger remains a cipher. I have more respect for this man than most I know or have heard of, yet I am at a loss to know what he thinks of me. Am I a friend? A colleague? A tool? All of the above? Or something that I cannot see but is clear to him. Perhaps I do not want to see myself reflected off his mirror, even with its evident tarnishes.

We work each day, cook well, talk, play, and a week rapidly passes. On our last full day together, we knock off work. Just as in the Karoo, we've been going at it all day, every day, and I call for a holiday. We decided to visit Mount St. Helens, the volcano that exploded in Washington State in 1980. It became one of the most moving experiences of our long relationship together, a metaphor for catastrophe and sudden death.

Odd how a resident of a place will somehow miss the usual visitor attractions. Just as I know many Parisians who have never taken the time to climb the Eiffel Tower, so, too, had I missed visiting the Volcano Observatory and other tourist points surrounding the now ruined husk of a mountain that had been the graceful, snow-covered cone of Mount St. Helens prior to May 1980. I loaded up my family and Roger in our aging Subaru wagon, and we made the two-hour drive from Seattle to the highway turnoff that leads to the mountain. As usual for this time of year, the weather was cool and rainy, and at the first visitor center, well down in the Cowlitz River Valley, a mountain weather-report station advised us that visibility near the crater was very limited. Yet we had come this far. Why stop now?

We headed east and began to gain altitude. Clouds obscured any long-distance view of the mountain, but as the miles passed and we entered the foothills leading to the mountain, we were soon enough surrounded by the effects of the catastrophic eruption. Roger was first to point out the obvious. We had stopped on a high lookout over the Toutle River, a watercourse with its headwaters on the fatal mountain itself. Prior to the eruption, this river had been a rather languorous and snaking flow within a wide valley, peacefully meandering seaward with its annual load of meltwater from the mountain. Now all that is gone. The river valley is still wide, but it is filled with rock and numerous small streams. There is no longer a single river channel, but now a series of braided rivers. A meandering river had been turned into a series of braided streams. Just like in the Karoo, after the Permian extinction.

Roger looked in wonder at this series of small streams where a large river once ran. He had lived so long in the Karoo and for so much of his life had studied the geological residue left by rivers both large and small. Yet here, before him, was the embodiment of those ancient Karoo rivers. The eruption of Mount St. Helens had stripped away most of the trees that had existed here up until 1980, and in so doing caused sudden changes in the pattern of sedimentation and then, ultimately, in the nature of the river itself. Immediately after the eruption, untold tons of new sediment and volcanic rock were carried into the river valley. At the same time, most of the vegetation was extirpated, either killed by the environmental effects of the eruption or smothered in the tons of suddenly arriving muck and sediment. The result was that the Toutle River had changed from meandering to braided, just as all the rivers did in the Karoo, and perhaps all the rivers on earth, at the Permian-extinction boundary.

We stopped for a picnic, my three-year-old son gamboling in the nearby grass, a vision of the earliest Triassic spread out below us under a steel-gray cloud cover. This river system will slowly return to its meandering ways as plants invade, tall trees grow, and the sediment washes away. It will recover, just as the world did in the Triassic. But Roger and I, and this little son of mine before me, all of us will be as dead as the Fouche family, long dead, before that happens.

After our picnic we continued, rising rapidly now into the higher altitude, and though we knew that the mountain must be there before us it was still entirely obscured. It was cold when we finally pulled into the David Johnston Lookout and Visitor Center on a ridge overlooking the destroyed crater itself. We walked the path, and then the huge panorama of the shattered crater was finally visible before us. It was overwhelming. It was a place of death, a place that I realized I had been avoiding for twenty years.

A nice metaphor, I thought. For I was convinced that we had long been standing before the mountain that was the Permian extinction without knowing how near to us it was. We had been straining to look far back in time, when we really needed simply to look all around us, in this world, even on this mountain.

Such thoughts were put off for the moment as we entered the David Johnston Observatory, for I needed to go mourn a long-dead friend. Just after the eruption, I hurriedly came to Seattle from my then residence in California. A week after the initial blast, I had talked a pilot friend of mine into flying me over the still-smoking crater. For an hour we crossed the terrain, saw the huge forests flattened and leveled, looked at a world without a trace of green, photographed the rivers gone from stately to chaotic, wheeled over the ash-covered slopes, flew over the graves of the scores killed by the sudden blast of that early May morning. And we flew as close as we dared to the grave of my friend, David Johnston.

David and I had entered graduate school together in the fall of 1971, two young geologists wanting to succeed in science and wondering if we had what was necessary to do that. We palled around together, took classes together. After two years we went our separate ways at different universities. It was a shock when I found that he had been the closest human to the mountain and the first killed by the blast. Did he suffer? I walked up to one of the rangers and asked if they had any pictures of David. They had but one, and the face was just as I remembered—unlined, a lop-sided grin, angular cheekbones, a mop of blond hair. I felt a yank of sadness, for his death, for my lost youth.

Roger and I looked now from the place were David died, where a huge concrete building had grown on a spot that once held his truck and gear, where his cells were bombarded by heat and energy. We looked down at a place completely different from the world prior to the eruption, a new world in the aftermath of destruction.

The huge blasted mountain was the very image of evil. There was no beauty here. A squatting abscess, a giant sore broken open and occasionally still oozing its poison onto the land, and more constantly into the air. Wrenched rocks, piles of ash, the wounds of a planet, a monstrosity. No plants, no life, just bare rock. And surely the thought arose as I look at this Hadean scene: *The aftermath of the Permian-extinction event would not have looked very dissimilar.* Even as we watched, the huge wrecked cone suddenly belched great roils of smoke, ash, and vapor into the air, a long blue plume of venom merging into earth's far more pristine atmosphere. This is what once killed a world, gas like this, the same gas now emerging from Mount St. Helens, but in far greater volumes. As we walked back to the car, a line of tour buses began to move out as well, and they, too, belched gouts of gray gas into the sky. We tried to walk by, and my little boy choked in the noxious exhaust, a gas not so different from that being belched by the nearby, ugly gash in the earth that had killed my friend and so many others twenty years before.

In silence we climbed into our car and did our part in further polluting this planet on the long ride home. During that journey back to Seattle, Roger talked of his life, a conversation that I recorded, one that became the heart of this book.

On our last morning together, Roger took photographs of my house and family. My young son frolicked in the garden; my wife looked on in morning dress. A perfect spring day, like so many we shared over the years in the Karoo. I thought of the Tweefontein farmhouse, once a place as vibrant as my own house on this day.

Scratched in stone in and around our house, as in that faraway house in the Karoo, are initials from my two sons, made at various times in their lives. Will some future time traveler come across these and wonder who these boys were and how they eventually died, just as I reflect upon the Fouche family and the long-ago family of mammal-like reptiles?

▪ ▪ ▪

The wheel of the seasons had again come full circle. I was newly forty when I first arrived in the Karoo in 1991, and in 2001 newly fifty now. Fall in the Northern Hemisphere, spring in the Karoo; the time for fieldwork. In October 2000, I once again flew the vast distance to Africa, once again rejoined the Karoo paleontology crew.

As usual, Roger was busy and successful. High in the hills north of the Bethel Canyon, he found a wonder: Just beneath the Permian-extinction boundary, he discovered a fossilized trackway site. Walking up a hill, Roger had spotted a single footprint on the surface of a small sedimentary bed. Digging back into the shale bank, he found several more. In his usual fashion, taking up where Hercules left off in his labors, Roger and his crew proceeded to remove part of the hillside to expose the trackway-bearing bedding plane over a several-day period. In the merciless Karoo sun, Paul October moved tons of rock. Not figurative tons of rock, *literal* tons of rock. The fruit of this labor was a cleaned sheet of rock, twenty feet on a side, a single sedimentary bed. It was once a shallow pond or a swamp beside one of the small rivers or streams. It was deposited but a few years before the extinction would sweep over the land.

A herd of mammal-like reptiles of several varieties had walked across this pond bottom. There were several sizes, within one of the species. The footprints were quite distinctive: five large and spreading toes, deeply sunk into the mud as would occur for a large animal moving across a soft surface, the smaller prints less pronounced and distinct. They are almost certainly from *Dicynodon*, the most common of the mammal-like reptiles at the end of the Permian. The distribution of sizes leads to intriguing speculation: Was this a family group, a herd? The footprints go in a straight line, suggesting a group of animals that were heading somewhere in determined fashion, not milling around, not stopping for a drink. Elsewhere on the huge sheet of rock, there are distinct impressions of smaller footprints. Some of these are almost certainly from cynodonts, our immediate ancestors. *Dicynodon* and so many other species would die out, in the event recorded only slightly higher in the sedimentary succession here. Luckily for us, the cynodonts would not.

Footprints bring back the past better than does any other clue, certainly better than bones. Here was proof of life, not a marker of death. Here was a snapshot from right before the extinction, the last days of empire; a photo preserved in sediment of movement; a reality, not a promise but a proof that there were indeed worlds and worlds more before our own. Here was *proof* of the Age of Protomammals in full flower, of diversity and life; not a few stragglers and last survivors but a splendor of life; not a remainder of it just before its fall. Were conditions deteriorating, with a sure prophecy of coming extinction apparent even to these small-brained animals? The record on this site does not argue thus. Or were these still the best of times in a warming world, with no clue to the tragedy just looming? The footprints offer tantalizing clues that the latter was true. There were many footprints recorded here from many kinds of animals—not a scene from a lonesome desert home to a last survivor but a scene of life, an afternoon around the porch of the Fouche farmhouse, just before whatever disaster befell that Karoo family, as a child and friends scratched their own proof of life into the flagstones lining the Fouche porch, not knowing that a few short years hence the heart of the family would lie cold and still in their desert graves.

I arrived at the footprint site a week after everyone else; I missed the excitement of finding these youngest records of Permian promenades and of other treasures teased from the surrounding rock. For it was not only news of footprints that awaited me: Roger Smith had already made other spectacular finds. Just beneath the boundary, he found two skulls of a crocodile-like animal in latest Permian rocks—an animal previously unknown from that time. After nearly two centuries of collecting, to find something totally new to science is almost unheard of. But who could ever underestimate Roger Smith? I arrived, listened to the catalog of new and impressive finds, and simply grinned and shook my head. More ticks for our chart, more proof of cataclysm.

With my arrival Roger took a new course of action. Not content simply to *find* the youngest record of footprints of Permian age from anywhere in the world, he proposed to take them back to Cape Town with him. Not literally; the stone slab on which they rested weighed more tons than even Roger Smith could ever manage. But clever Roger had another plan.

On the last day of this trip, we drove to the site of the footprints. Roger had many hundreds of pounds of plaster in the trucks, and equal weights of water. He intended to make a plaster impression of the footprints and bring that back. The morning was all about manual labor: shoulder a hundredweight sack of plaster, struggle up the hill until the heartbeat hit the wall—130, 140 beats a minute, 160, who knew?—heart smashing in the chest, sweat streaming, shoulder carrying the great sack squeezing onto the rest of the body, backbone compressed, resting and panting, and on up the hill. Sack after sack. And then back to the truck, the huge carboys of water next, the same trip. Paul October, Roger Smith, Hedi Stummer, and myself, old friends, longtime coworkers, nearing the end of a long and profitable association, all growing older, all racking our bodies in the African heat with this donkey's job. Not for the first time, I reflected on our collective efforts and wondered about us, candidates for the couch, one and all.

For most of the day, the mad plasterers were at it. First coating the fossils with grease so that the plaster would not stick, then applying layer after layer of plaster and burlap on top of the rock surface. I was not needed in this process, so I explored around the site. We were high on a hill, and the panorama around me was both familiar and new. The Bethel Canyon was off to our right, and from my vantage point I could see the broad and muddy Caledon River. The game park was visible on the other side, and the view was a portal to multitudes of memories, each a time and given day, a lost reality, somehow encapsulated in blood and flesh and electricity, or whatever memories really are made of. The memories came, and they dredged up another electrochemical phenomenon, as I suppose it to be—emotion. For the memories were not free, and while many liberated joy and satisfaction, some arrived only with a cost of pain, the searing pain and despair of lost friends and lost times as the price of admission, and the knowledge that this long time in the Karoo was about over for me, that these companions were about to become the past for me. How does a paleontologist deal with pain? There's always the escape of the search, the quieting of the mind necessary to find anew and to create new and better stores of memories. So I searched.

There was bone about, and I found an occasional fossil. I wandered across the side of the hill, lost in memory, lost in rock. Up into the Triassic rocks now, across the boundary that so stubbornly gave up its secrets, I felt another emotion: contentment in the huge efforts and solid victories of discovery. But I also played the "what ifs" a bit, trying to measure the gains with the losses. And while I was deep in thought, my finally well trained eye registered and alerted me to the presence of an eggshell on the ground ahead, sitting in deepest-red strata proclaiming the Triassic Period, the time just after the mass extinction. I bent down, knees popping at the insult, and looked at the "egg." A new insult now. I could not see it up close. I reached into my shorts pocket and pulled out a pair of reading glasses, a new *tempus fugit* message that my body had now given me, eyes that could not see anything up close, eyes that had never needed correction before, a new reminder that the time of my life was passing, and much had already passed. Another reminder that somewhere ahead death awaits me, as it does for us all.

Now bespectacled, I peered at this round object. It was bone, an obvious braincase. Whoosh. A hit of joy, I snarled a grin— *gotcha!*

I was looking at a distant ancestor. A cynodont, the Permian-extinction survivor that gave rise to us.

I had a sudden, distant memory from teen days, a snatch of song about saving the last dance for me. I knew that this would be my last find in the Karoo. And just as at the gym in my far-off home, where my best friend and longtime workout partner would never let me leave without a last swish of the basketball, so, too, was I going out of the Karoo with a hit.

Later in the day, I got Roger to saunter over, and he confirmed the identity of this small skull. It was quickly plastered and then extracted. I got the barest of nods from Roger, and that was reward aplenty.

The day was moving toward dusk with a speed no longer surprising. The sun mocked us by moving so fast, and Africa tired of its daytime raiment and readied for its dark night. Roger gathered us around the white rectangle. It was the size of a door, and he'd brought a real door to attach it to, since the plaster by itself would easily shatter at any of the many jolts and insults it would certainly receive on the long trip back to Cape Town. We were instructed to gently put our hammers all around the edges and simultaneously pry upward. We pried, but nothing happened. The huge plaster sheet was wedded to the stone, seemingly irrevocably. Another shove on Roger's command, and again nothing but increasingly strained backs. Intractable. But I'd known Roger for far too long to put any sort of smart bet on the plaster's staying in this spot forever. It would be removed, and probably soon. Roger stared at the recalcitrant object, squinted thoughtfully, and then glanced about. A large hardwood bough was found and levered beneath the plaster. A cry, a loud crack, and the plaster was free. Roger smiled triumphantly.

We all put our fingers under the now freed plaster, and on "three" we lifted with gusto. The plaster sheet was heavy, terribly heavy, and we strained to bring it up on its side. Up it came, into the setting strands of African sun, flies meandering like air plankton about this great rectangle of white. It now seemed like the non-evil twin of the monolith from *2001: A Space Odyssey*. The door was slid underneath, and the plaster gently dropped down and strapped to it. I shook my head. The mad race of paleontologists.

And now for the next trick, as they say, but here there would be no sleight of hand. We had to get the door down to the waiting truck. There were four of us here. One of us was Hedi.

She comprehended the next act and immediately told Roger that she would not be able to lift it. Hedi is terrified of barbed-wire fences, and we would have to go over two to get to the car, a quarter mile away and at the base of the steep hill we were on. We would have to hump this weight over the rocky Karoo terrain. We would need her. I had no doubt that this strong and deep woman could do this, if only she would let herself. But she quailed, sought excuses, looked at the terrain, and balked.

Roger talked to Hedi. He told her she could do it. She was very doubtful. Each of us took a corner of the door, Paul and myself in front, Roger and Hedi in back. We heaved it up onto our shoulders and began our trek. The weight was dreadful; I could feel my knees compressing and my back complaining, sweat popping free, and we had gone but a dozen feet. Hedi cried out that she could not do this and wanted to drop her corner, but she could not, or the plaster would be lost. Somehow we struggled forward.

We rested every thirty feet or so, dropping the door onto its side. Time passed. Lift, carry, rest; lift, carry, rest. We got the great burden to the first fence, and Paul and I jumped over with the door resting on a strand of barbed wire, the whole fence groaning and threatening to collapse. The others pushed when we were in position, and we slowly slid it through the wires. Lift, carry, rest, to the next fence, down the great hill. What did we look like to the tribes walking past us on the road below, our goal? What did these four mad people look like to those tribes of Africans slowly going from A to B when we were in the middle of nowhere, far from any defined A *or* B?

The car was in sight now, and darkness was rapidly gaining on us. We had to get this done. We were all soaked in sweat. Lift, carry, rest, and we eventually got to the truck. The last hoist was the hardest, one last effort, and then all was done.

Hedi was disbelieving. We looked at her with quiet satisfaction, and for the first time, at least in my experience with her in the field, she seemed to view herself in the same way. Good job. Our last job together.

The plaster and the crew decamped for Cape Town the next

morning, and I took my leave before Roger loaded up and headed off. I found my rental car and started the long road toward home, family, and safety. There were two Cape cobras on the dirt road that was the first step back toward the North American civilization that I am part of, but they were already part of a new past in my life even as I rolled by their brilliant tan coils. I had one last stop to make.

On the way back to Cape Town, I paid my last respects to the Fouches. Standing by their grave, I wondered at all the Karoo nights and days, at what we had learned. There was a certainty that framed the hard work of Roger Smith and our crew over the years of our project. We knew now with confidence that the extinction on land was fast—just as Doug Erwin, Sam Bowring, and their Chinese colleagues had found that it was equally fast in the sea. All the years in the field—the joy, hardship, pain, and pleasure—had resulted in a single truth: The extinction was rapid, a catastrophe tumbling down on a world with alacrity and utterly changing it.

But what we did not know—still, after all this work—was the cause. This was a source of enormous frustration. We geologists had brought our best efforts to the table and walked away still mystified. Surely such a killer could not go unidentified. Or could it? And then, in one fell swoop, a new methodology and a new term seemingly did away with all doubt as to cause. A completely different field of scientific study seemingly grabbed the prize by looking for—and finding—buckyballs.

Chapter 11

Buckyballs

JUNE 2000

Every four years or so, those interested in the study of mass extinctions—the geologists, geochemists, paleontologists, astronomers, and oceanographers, mainly—gather in what has become known as a "Snowbird" conference, named after the site of the first two conferences held in the 1980s in Snowbird, Utah. Each of these conferences has dealt with mass extinction, and not surprisingly, since their inception soon after the Alvarez discovery was announced in 1980, each has been dominated by scientific presentations dealing with the Cretaceous/Tertiary extinction. In the summer of 2000 the site of the new "Snowbird" was Vienna. But this meeting, while having a great deal to do with the K/T extinction, was also a passing of the guard. The Permian extinction was the star attraction—as evidenced by the work being done by Walter Alvarez himself.

Many of us described our research into the Permian-extinction event, including Mike Rampino, Sam Bowring, Henk

Vissher, and, most important, the dean of Permian-extinction workers, Douglas Erwin. Each talk arrived at a similar conclusion. No matter what group of organisms was studied, no matter where in the world the study was carried out, the extinction was found to have been a rapid and catastrophic event. In this there was no disagreement. But the cause? There lay the frustration for all of us, for unlike the K/T event, which left abundant and varied clues for us to find, the Permian-extinction killer was a far more silent and skillful assassin.

Doug Erwin's summary of the event was especially instructive, and at the same time it told the tale of frustrations we all felt at being unable to pin down the cause. Some years earlier Erwin had championed a theory that he called "The Murder on the Orient Express Explanation": that just as there was no single killer in the great Agatha Christie whodunit, so, too, was the Permian-extinction event really the end result of the earth's undergoing a multitude of stresses that, when combined, caused the hideous mass extinction. But here in 2000 Erwin came to a new view. His work with Sam Bowring on the Chinese sections had shown that the event must have taken place in 165,000 years or less, with emphasis on the "or less."

Erwin presented a summary of his many years of fieldwork that had included both Chinese and American scholars. This work, published in *Science Magazine*, mirrored our own results from the Karoo, but with far more data than the few fossils we had so laboriously collected over a decade of work. The China effort combined results from five different stratigraphic sections in the Meishan locality, with sampling intervals made every thirty to fifty centimeters. A total of 333 species of marine life was ultimately found in these rocks, including such varied sea creatures as corals, bivalve and brachiopod shellfish, snails, cephalopods, and trilobites, among others. In the last meter of strata deposited at the very end of the Permian Period, Erwin and his colleagues managed to find and identify a whopping 172 different species of fossils. Nowhere at any stratigraphic horizon at any time has so thorough a collecting effort—or so rich a fauna—been documented with such precision. The death toll in this meter of strata is also staggering—94 percent of the species suddenly disappeared.

Like all of us, Erwin searched for cause among the many threads of evidence left behind in the rock record. The various environmental conditions in the seas at the end of the Permian included widespread evidence of oceanic anoxia, or low oxygenation of seawater, in both the shallow and deep seas. The anoxia was apparently of such magnitude that many marine organisms were rather suddenly killed off, just as they are today in modern red tides. There is also evidence of global warming at the time of the extinction, and the coincidence—if that is what it was—of the Siberian lava eruptions at the same time as the mass extinction. How to account for these various lines of evidence, and how could they add up to a possible single cause? Erwin summarized the various suspects. First is the possibility that the Siberian traps, or volcanic outflow, introduced huge volumes of gas into the atmosphere, triggering large-scale climate change and acid rain, as earlier suggested by Paul Renne and others. With new information from disparate sources, a sudden methane release into the atmosphere became a viable candidate for the killer. But other mechanisms could not be ruled out, according to Erwin, including everyone's favorite at this conference, an asteroid or comet strike. Though there was only the slightest controversial evidence, the rapidity of the Permian-extinction event argued for some sort of "quick strike." Among potential causes of mass extinction, only asteroid impact is known to be capable of such destruction in so short a time.

Nagging doubts remained. If an impact caused this extinction, where were the well-known clues so common amid the last layers of the much younger dinosaur extinction, the 65-million-year-old K/T event that Walter Alvarez had studied? Yet if there was not a smoking gun pointing to impact found in the boundary layers in China, there *was* the discovery of a new crater.

The conference was abuzz with gossip about a purported large impact crater newly discovered in the Woodleigh Carnarvon Basin of western Australia. With a diameter of 120 kilometers (about 60 to 80 kilometers less than the Chicxulub Crater in Mexico that caused the K/T extinction), this impact would not have been as catastrophic to the biosphere as the K/T event, but a huge blow to the world nevertheless and surely capable of causing mass extinction. The unfortunate aspect of this crater was that its age was very poorly constrained. Nevertheless, the authors of the paper describing this new discovery trumpeted it as having been caused by the true Permian-extinction killer, in their mind a giant comet or asteroid. For many of us, the nagging doubts lingered. The lack of supporting evidence from the many Permian-extinction boundary sections was the most disturbing aspect. The K/T impactor left a diagnostic signature. If the Woodleigh crater was a Permian-extinction impact crater, it was far more subtle in its effects.

Many lines of evidence were converging on something more prolonged than a single quick strike. In September 2000, University of Oregon geologists Evelyn Krull and Greg Retallack published a paper detailing their results from prolonged geological and geochemical studies of Permian-extinction boundary sections in Antarctica. Their results strongly supported the idea that the early Triassic was a time of heightened methane-gas volumes in the atmosphere. Methane is one of the most potent of the "greenhouse gases," and its sudden release would have driven global temperatures sharply higher. These results followed on Retallack's 1999 findings from the Sydney Basin in Australia. There, Retallack recognized that the Permian-extinction boundary was coincident with the formation of the last coals anywhere on earth for many millions of years of Triassic time. The boundary coincided with a large-scale extinction among plant species as well as a dramatic changeover in climate, as deduced from fossil flora and fossil soils. The latest Permian of Australia was characterized by deciduous flora adapted for a humid but cold temperate climate. At that time Australia, like nearby southern Africa, was located far nearer the poles than the equator. In the earliest Triassic, however, a marked change in climate apparently occurred. The fossil-soil types indicate a much warmer climate—as would occur from a sudden onset of global warming. Coal formation abruptly ceased. Sedimentation rates markedly increased in the lower Triassic rocks, and, as we did in our study in the Karoo, Retallack interpreted this as the result of extensive and sudden deforestation at the Permian-extinction boundary.

Other suggestions of a profound world-changing event came from Roger Buick, a geoscientist from Australia. Buick, a specialist on the Precambrian world, became intrigued with the Permian-extinction event because of how it sent our world, for a short time, back to conditions quite like those prior to the rise of complex animals and plants. Buick described the event in Australia as having been caused by some sort of repeated environmental perturbation, such as the pulsed release of greenhouse gases, the repetitive overturn of a stratified ocean, persistent prodigious volcanic exhalations, or a succession of comet impacts. The key was that it was not a single shock, as happened at the end of the Cretaceous. None of the observed evidence suggested a single asteroid impact. A succession of things happened.

A sense of frustration was settling in. The sum of geological, geochemical, and paleontological data gathered by so many scientists for so many years remained inconclusive. One thing was certain: The body count at the end of the Permian had been high, and the extinction had been fast acting. And then, as the year 2000 gave way to Arthur C. Clarke's year of 2001, a much-trumpeted article appearing in a February 2001 issue of *Science Magazine* called the controversy over: They had found the "smoking gun"—and proof that the greatest mass extinction in history was caused by the collision of a titanic comet with the earth. There was personal irony in this for me. The research team announcing this major discovery was headed by scientist then at my own university, a woman named Luann Becker. With great flourish at a NASA press conference convened for the occasion, Luann imprinted a new term on the lexicon of public consciousness concerning things scientific and catastrophic: buckyballs.

In this paper Becker claimed that the Permian-extinction event was entirely caused by extraterrestrial impact. While unable to pinpoint the site of the impact, Becker and her coauthors asserted that the space body had left a calling card—a much higher level of complex carbon molecules called buckminsterfullerenes, or buckyballs, with the chemically nonreactive gases helium and argon trapped inside their cage structures. Fullerenes, which contain at least sixty carbon atoms and have a structure resembling a soccer ball or a geodesic dome, are named for Buckminster Fuller, who invented the geodesic dome.

Becker suggested that the gas-laden fullerenes were formed outside the solar system and that their concentration at the Permian-Triassic boundary could mean only they were delivered by a comet or an asteroid. The researchers further estimated that the comet or asteroid was six to twelve kilometers across, or about the size of the K/T impactor. The article went on to say that telltale fullerenes containing helium and argon were extracted from sites in Japan, China, and Hungary, where the sedimentary layer at the boundary between the Permian and Triassic periods had been exposed. The evidence was not as strong from the Hungary site, possibly because the sample came from slightly above or below the boundary layer, but the China and Japan samples bore strong evidence. Fullerenes were supposedly found at very low concentrations above and below the boundary layer, but they were found in unusually high concentrations at the time of the extinction.

For Luann and her team, it was a triumph. The paper—and the discovery—was the stuff that glittering careers are made of. NASA beamed with pride, for NASA had funded the bulk of the work, and it was NASA that flew Luann to Washington to announce the results soon after publication of the research paper in *Science*. (To me this alone was a point of wonder. For years all of the mass-extinction work had been funded by the National Science Foundation. But with the advent of the new century, NASA became increasingly interested in the link between impact and extinction.) It was one of those stories that the press loves: death, catastrophe, and a new angle, for the buckyballs, with their contained atoms of helium 3, made wonderful graphics and compelling copy. Soon the story was found in newspapers and magazines around the world. A major mystery apparently solved.

But for those of us who had slogged for years in the killing fields left behind by the K/T catastrophe, there was a sense of doubt. While there was a slight yet unmistakable feel of sour grapes by so many of us at getting scooped by this research team, there was also a feeling that not everything added up. For instance, one of the lead paragraphs of the press release— "The collision wasn't directly responsible for the extinction but rather triggered a series of events, such as massive volcanism and changes in ocean oxygen, sea level and climate"—made no sense at all. How could a comet impact create volcanism or sea-level change? Much was known about what large-body impact on the earth could or could not do, and this was in the realm of the "could not do." Yet much of this could have been zeal on the part of the press rather than the Gospel According to Luann. So those of us in the wings held our tongues. For a short while anyway.

I had first met Luann Becker the year before, at an astrobiology conference, where she had presented her preliminary findings on buckyballs and noble gases from the Permian/Triassic boundary sections. At that time I concluded that she was a good chemist applying a whole new technique to the study of extinction—always a good thing—but that she was somewhat naïve about important geological and biological matters. Nevertheless, how could I not be intrigued? Because of this I managed to get her a one-year-sabbatical replacement position at my university. I wanted to learn more. She agreed, and in the summer of 2000 she moved to Seattle and brought her entire lab with her.

During the fall of that year, she showed me the first draft of what would later be her paper. It was clearly a contribution of importance, and the hoped-for culmination of the quest so many of us had pursued for so long. But the end of the paper held several breathtaking flights of hyperbole concerning the undoubted role of an asteroid impact as the sole cause of the extinction. She also introduced aspects of of planetary geology that seemed to have nothing whatsoever to do the with the rest of the paper. I suggested that some of these might be more cautiously worded, for so many aspects of the P/T extinction did *not* look like the asteroid-caused K/T event, our only yardstick on how the world and its biota responded to a large-scale impact. Finally I suggested that the paper could be strengthened if some way was found to use her measurements to arrive at an estimate of the impactor's size—was it larger or smaller than the ten-kilometer K/T impactor? I scribbled these criticisms onto the preprint of her paper and gave it back to her, expecting (and hoping for) a lively give-and-take. Though my suggestions were followed, my comments were received in uneasy silence. It was to be the last time that she ever sought my advice or even discussed the extinction in any way with me.

It's safe to say that the formal publication of Luann and her coauthors' paper in *Science* did not go unnoticed among those who had long toiled in the extinction game—and in studying the Permian-extinction event in particular. The wise ones gathered and talked it out and were left perplexed. I certainly was. By this time Roger Buick of Australia had been hired away from his home university in Sydney to take a position in my department, and I could not help but remember his comments about the Permian-extinction event, based on his decades of observation of rocks from this time period exposed in Australia: "Clearly, a single impact could not have been responsible. The most obvious interpretations are repeated environmental perturbations, such as methane hydrate melting pulses, repetitive overturn of a stratified ocean and/or persistent prodigious volcanic exhalations, or *serial* extra-terrestrial insults." This view certainly jibed with my observations on the South African rocks, as well as with my long years of looking at Cretaceous-aged mass extinction, where the pattern of fossils and rock exposures at the mass-extinction boundary *did* look as if they were caused by a single catastrophic asteroid or comet impact.

Several days after the publication, I received an extraordinary e-mail from my close friend and colleague Joe Kirschvink, who had shared so many Karoo adventures with me. By happenstance two of the major experts on the Permian extinction, Doug Erwin of the Smithsonian and K. Isozaki of Japan, met with Kirschvink and Ken Farley, the world authority on helium, to discuss the new results. They were curious about one of the statements about the fullerene concentrations above and below the boundary layer.

In reality, the Hungary site showed *no* evidence of fullerenes, so the critical evidence came from the other two sites. Both of these suites of rocks were intimately familiar to the assembled group at Caltech, for Doug Erwin and his colleague Sam Bowring had collected the Chinese samples analyzed by Luann, while Isozaki had done the seminal work on the Japanese Permian-extinction boundary that Luann and her group had analyzed. Yet, unknown to Becker and her colleagues, the Japanese samples had not come from rocks at the Permian/Triassic boundary. The Japanese samples, which had been forwarded to Becker by her colleague Mike Rampino and a Japanese geologist, had mistakenly been collected from rocks millions of years younger than the 250-million-year-old Permian/Triassic boundary. That left only the samples from China as proof of an impact. It also left a huge residue of unease among the Permian-extinction specialists.

All science is predicated on replicability. Ken Farley and his group had asked Luann Becker for splits of her critical Chinese samples. Luann replied that she had used up all her samples in the analysis and could supply no more. Farley then went to the original source of the samples, Sam Bowring. Sam promptly sent new material, which was duly analyzed both at Caltech by Farley and at my university by Randall Perry. Farley used a blind-sampling technique, asking Sam to withhold any information about which samples came from the critical level where Luann had found the helium-bearing buckyballs. After exhaustive tests, Farley was not able to replicate the Becker group's findings. There was no helium 3 to be found.

Speculations were rampant after this disappointment. The most likely reason for this negative finding was that the He_3 layer discovered by the Becker group was an extremely thin layer from a more massive sample collected and supplied by Sam Bowring of MIT, and that the material later sent to the Caltech group did not sample this exact bit of rock. There was also some speculation that the Becker findings might somehow have been related to lab error or contamination of her glassware in some fashion, although this seemed quite improbable. While not crack field geologists by any means, all of her group were excellent lab chemists, with long experience in the sorts of analyses in question.

By the year 2000 it was generally accepted that in all probability an impact *did* occur, some 250 million years ago. There was every reason to believe that such an impact would have caused some—perhaps much—of the chaos and species death that define the Permian extinction. But the question that remained to be answered was whether or not this fiery messenger from space was the sole assassin.

Chapter 12

A New Kind of Exinction

NOVEMBER 2001

By 2001 a long-anticipated consensus about the Permian extinction had been reached. The mystery seemed defanged. Not exactly *solved*, but now limited to just a few options. Not much except crashing asteroids could wreck a world and kill off so much in such a short space of time. Some monstrous event had indeed obliterated the Permian biosphere of planet Earth 250 million years ago. In China the exquisite work of Doug Erwin, Sam Bowring, and their Chinese colleagues had shown that a short period of extinction had eliminated most marine life in a period of a hundred thousand years or less. In the Karoo, Roger Smith and I had demonstrated that there had also been a relatively short period of catastrophe that wiped out the mammal-like reptiles. Finally there was evidence that these catastrophes, one in the sea and one on land, happened at the same time, for the isotope work conducted on Karoo nodules by Ken MacLeod suggested that the climax of the extinction was simultaneous on land

and in the sea. A certain smugness set in among the principals of this research. And why not? We had worked for many years, and now a satisfying end was in sight. And then all this wisdom was challenged by new findings.

The potential unraveling of the theory that the extinction of the mammal-like reptiles in the Karoo coincided with the extinction of the marine fossils in China came about in a strangely serendipitous way. In 1998 a geologist from Wyoming named Maureen Steiner had traveled to South Africa to attend a meeting dealing with Gondwana geology. Like Joe Kirschvink, Maureen studied magnetostratigraphy and was the world's resident expert on the pattern of geomagnetic reversals spanning the Permian/Triassic boundary. After the conference she took some oriented cores from Karoo rocks located about an hour's drive from Lootsberg Pass. Later she came back with one of the world's experts on the Permian extinction, Mike Rampino of New York. While their original goal of obtaining a valid pattern of magnetics across the boundary did not work because of excessive heating of the strata by the ever-present dolerites, they did gather a detailed sample of palynomorphs. These microscopic reproductive bodies of plants consist of tiny spores and pollen grains that are often preserved in sedimentary rocks. They sent their samples to the lab of specialist Yoram Eshet of the Israeli Geological Survey, and he succeeded for the first time in isolating these ancient plant fossils from late-Permian Karoo rocks. He found a continuous record of typical Permian plant fossils, followed by an interval of strata containing only the remains of fungal material, followed by fossils of typical Triassic plants. The fungal interval was interpreted as the (by now) well-known "fungal spike" that had been observed at various Permian/Triassic boundary sections around the world (including that in China, where the marine extinction was so well marked). Maureen Steiner, Rampino, Eshet, and a graduate student then wrote up a paper reporting the first recognition of the fungal spike in South Africa. The position of the fungal spike, in this single stratigraphic section first studied by Maureen, was within the first Katberg sandstone. The authors then deduced that this position in the Karoo coincided with the position of the marine mass extinction in China.

I received a preprint of the paper and was mystified. In the original version of the paper, it was unclear where their fungal spike was located. But the authors kindly sent me a digital photo of the position, a stratigraphic position that Roger Smith and I considered to be well in the Triassic, not at the extinction boundary of our mammal-like reptiles. And that was the rub. The lowest Katberg sand was some *thirty to forty meters* above the position of the mammal-like reptile extinction. The time necessary to accumulate thirty to forty meters of strata in the Karoo Basin would have been on the order of tens to hundreds of thousands of years. If this discovery could be confirmed, it meant that the great Permian extinction had at least two phases—first a phase onland, in which the mammal-like reptiles died out, and then a second phase, wherein the many invertebrates and vertebrates in the oceans were killed off. Never had such a mass extinction been observed, one that dealt death first by land, then by sea. This was a major discovery—if it could be confirmed. Unfortunately, the observations leading to this conclusion were based on data gathered on a single section.

At about this same time, in early 2001, another bit of evidence arrived that suggested that the Permian extinction was anything but a single-stroke event. Australian geologist Roger Buick, an expert on the origin and early history of life, had coincidentally been studying the nature of the Permian-extinction record in Australia. He had analyzed several stratigraphic sections from western Australia and found not just one isotopic perturbation in carbon 13 but at least seven in the short interval of time crossing the Permian/Triassic boundary. Such perturbations are not local but global effects. If there were this many perturbations in ancient, Permian Australia, there had to be the same number everywhere on the globe—including, of course, in the ancient Karoo. Yet the earlier work of Ken MacLeod suggested that there was but a single perturbation in carbon from the Karoo.

One extinction or two? One carbon perturbation or seven? How to choose? We had to go back, of course. And in October 2001 we did—a bunch of us this time.

▪ ▪ ▪

The Lootsberg Valley spread out beneath us like a great brown tablecloth. Enormous vultures rode the thermals, but otherwise the vista was exanimate. Only a thin ribbon of road brought order to the wide valley floor, and there was the sense that the living shared this place with the fossil dead. The air was so clear that great distances lay visible, as if the landscape were part of another, larger planet, where the very horizon receded impossibly far. Green faded to relentless brown as it fought a losing battle among the low, dispirited shrubs and thorny scrub. The world seemed to have encountered a great test and been found wanting. A test of life failed; failed in the present, failed immeasurably more so in the quarter-billion-year-old past.

I was perched on top of a cliff at Wapatsberg. Fanned out along the ravine below was a new crew, one that I brought from America to bolster the traditional South African team. Roger was there, but the rest of the famed Karoo paleontology team had drifted away. Their places were taken by two Australians now living in America: Roger Buick, newly hired at my university, and Greg Retallack, from the University of Oregon. Farther down the slope was a keen undergraduate from my school, Tom Evans, and ever farther afield was my son Nick, now eighteen. He was seven years old the first time I came to Africa and had grown from boy to man over the many years we'd been tailing the Permian killer.

We were not here to find fossils. Nothing even remotely as interesting as that. We were here solely to collect tiny samples of sediment and return them to our giant machines, to compare the results of isotopic tests of the Karoo with those from Australia and Antarctica. I had a premonition of what these results would show, and, as it turns out my premonition was correct. In October 2002 the results began to come back, showing that there was not a single isotopic perturbation, but several. Enough to tell us that no single asteroid or comet caused this extincion or that the earth at that time was hit at all.

It seemed like a mass of contradictions. At first anyway. But science is often about disproving things, not proving them. And so we began by showing what did *not* cause the Permian extinction.

First of all, it was not a long, slow event, as geologists had believed for decades. But, more surprisingly for those of us who worked on the K/T extinction of 65 million years ago—an extinction clearly caused by a single rapid catastrophe—it was not a rapid event either. Neither long, continuous, and slow nor short, swift, and of single cause—the Permian extinction unfolded as a series of short but successive events.

The record of the extinction in marine sections did indeed seem to be a single short event. The wonderful work of Doug Erwin, Sam Bowring, and their Chinese colleagues went a long way in proving that. That type of extinction could be caused by a single, large, devastating asteroid impact with the earth. But our Karoo work showed repeated devastation. It looked like multiple impacts, if impact were the cause at all.

And so a picture emerged of something new, a type of extinction never envisioned before. Fast, and in pulses. Not a single short burst of death, as at the end of the Cretaceous, but instead a series of episodes of extinction, one after the other, for perhaps a hundred thousand years. We still do not know the cause, but we know the duration and nature of the killing now. It is a new kind of catastrophe, and it implies new directions in research we are only now beginning to grasp.

The supposed cause of the Permian extinction thus seemed to be a moving target, decade by decade—from climate change to something due to volcanism to meteor impact to something related to methane in the atmosphere. This is how revolutionary science often works. Quite often a project deemed completed has a way of mutating into something far from finished and often far more interesting than originally thought. Such has been the case of the Permian mass extinction. Until 2002, I despaired of ever knowing its cause. And then, as so often happens, new results allowed new connections. The killer, it turned out, was hiding in plain sight, in the most obvious aspect of the late Permian and Triassic rocks in the Karoo.

Chapter 13

Resolution

There were many days when the rocky Karoo yielded none of its treasures. Of course, we bone hunters took these days personally. Like baseball players searching to find a lost stroke, all of us suffering through the inevitable days when we could find no scrap of bone, no hint of a fossil, would try to rekindle our skill and luck. On such days, when the frustration of hard looking without reward became too oppressive to bear, there was always a fast solution. Behind Roger Smith's back, we would furtively traverse into higher strata, leaving the often barren, uppermost Permian strata and trespassing up into the forbidden riches of the lowest Triassic.

For reasons still obscure, these lowest Triassic rocks are extraordinarily rich in fossil, much of it quite spectacular. The problem was that almost every one of these fossils came from but a single taxon, the ubiquitous dicynodont *Lystrosaurus*, survivor of the Upper Permian. I would always rationalize these illicit jaunts as trying to "get my eye in"—to find a fossil bone or two just to remind myself what to look for. I was not alone in this behavior, for I would sometimes find another sheepish member of our crew similarly engaged. Roger would catch us in these forbidden beds and angrily shoo us back down into the Permian. Such was our life—we lived and searched in the highest beds of the Permian and only occasionally had glimpses of the aftermath of the great dying. There was a paradox here. Even if fossils were more abundant in the Lower Triassic beds, their diversity was not. Over the years in excess of fifty genera of vertebrates have been recovered from the highest Permian beds. Yet fewer than a tenth of that number can be found in the succeeding Triassic beds.

Could these numbers simply be an artifact of sampling or of the vagaries of fossil preservation? Perhaps it could be argued that the so-called catastrophe affecting the composition of the vertebrate faunas was not due to an extinction at all, but related instead to a change of environmental conditions that favored the preservation of fossils in the Upper Permian compared to Lower Triassic rocks. Yet clearly this argument is easily refuted, if necessarily made (all good science requires a null hypothesis). Because fossils are more common in the Triassic side of the Permian-extinction boundary than in the Permian, we would expect to find as many or more types of fossils (taxa) in the Triassic. But such is not the case. In spite of a rich fossil record, the Lower Triassic was a time of low diversity.

The transition was so striking, the patterns of abundance so stark, that I hungered to see higher in the Triassic, to better witness the long recovery of the land following this most devastating extinction. But our work never allowed this. We would make quick visits into the Triassic and then settle back down into the Permian once again, amid the richer world of dicynodonts and Gorgons.

In a larger sense, this pattern also carried for the long road trips into the Karoo. With every expedition I would start in Cape Town and then drive into the interior. Because of the geological context of the bedding, these drives would be voyages up through time, beginning in the 400-million-year-old Ordovician rocks near Cape Town. The farther I traveled from Cape Town, the higher into the sedimentary succession I would find myself. The apogee of these voyages would always be near the small, central Karoo town of Bethulie, the hamlet situated right at the limit of the youngest Permian and oldest Triassic beds. From this point I would look wistfully northward, where even younger strata could be seen stacked into distant hills. I longed to break free from my prescribed orbit, to visit these younger times. But invariably my trajectory would again fall back toward the south, and Cape Town. Finally, at the end of my last Karoo trip, I broke free and took a long car trip north across South Africa, crossing a country and a time interval simultaneously. On this trip I traversed the entire Triassic system, from the oldest rocks at the P/T boundary to the top—marked by another mass extinction, that at the end of the Triassic. This was a journey through 50 million years of rock.

Leaving the familiar areas around Graaff-Reinet and Bethulie, I gradually climbed higher in the sedimentary succession of the Triassic, moving well into the African escarpment and skirting the plateau kingdom of Lesotho. Virtually all of it was bright red in color, a variegated landscape boldly painted in an infinity of those shades belonging to the family of red. Crimson, scarlet, ocher, orange, cherry, pink, ginger, burgundy, ruby—a restricted spectrum of sedimentary beds gleaming under the hard, hot African sunlight, reflections of blood and the blood shed by mass extinction.

As in the lowest rocks of the Triassic that we had so long sampled, fossils were not rare. *Lystrosaurus*, the staple of the lowest Triassic, could be found in its multitudes for another several thousand feet of strata above the Permian-extinction boundary beds. It must have been the most common larger vertebrate in the world following the catastrophe, and we can imagine the scene soon after the initial carnage of the extinction, a world where huge herds of this piglike animal were reestablished, among a few smaller mammal-like reptiles, including (thankfully for us and all mammals) some number of the cynodonts, rootstock of all mammals. But so much else was gone then: the herds of larger dicynodonts, the many small herbivores, the myriad carnivores—and the Gorgons, of course. The last of these are found in the highest Permian, and then no farther.

The time of the mammal-like reptiles did not end with the Permian extinction. The few that survived provided a rootstock for a new branch of their tree, and they even made a minor resurgence during the Triassic Period. But their dominion was over. They had one last moment of grace: Among the lineages of evolving cynodont survivors, one stock made the jump to the mammalian grade of evolution, with differentiated teeth, a secondary palate, the mammalian middle-ear configuration, and, we suspect (but cannot prove), those other all-important aspects of "mammalness"—body hair, warm-bloodedness, live birth, and the nursing of the young. Now, as I write this in 2003, I think I know why they survived.

I passed up through the Katberg Formation, rocks rich in *Lystrosaurus* fossils, then moved into the overlying Burgersdorp Formation, and then into higher rocks yet. By the latter parts of the Triassic, the true mammals were loose in the earth. But they were certainly not creatures striking fear into the rest of the world's animals. The first true mammals were small, with skulls only several inches long. They may have taken to the trees or down into the ground, and they may have lived like present-day shrews. Their path was that of the ignoble, a life of eating insects. They were a far cry from their (by then) distant ancestors, the mighty Gorgons of the Permian.

During the Triassic a whole new suite of vertebrates populated the land. The oldest true "ruling reptiles," the stock that would ultimately give rise to the dinosaurs, are found in the oldest rocks of the Triassic. The most ancient of these is *Proterosuchus.* Yet within several million years, there was a variety of these predinosaurs, belonging to a group known as the archosaurs. From these ancestral stocks the successful lines of crocodiles and crocodile-like animals known as phytosaurs evolved. From other members of this group came lizards, snakes, and, by the middle of the Triassic, the first true dinosaurs.

As one drives north, the Karoo eventually ends. There is no sudden boundary. The dry land and rocky vistas become more tamed; endless grasslands and neat farms replace the harder sheep ranches, and as one approaches Johannesburg, the Karoo is but a distant dream. About two hours from J'burg, the last outcrops of fossiliferous rock can be seen. The rocks here are so distinctive that they can be observed from miles away, for finally, after many hundreds of kilometers—and 50 million years of accumulated time—the Triassic comes to an end. It is overlain by a new color, a dazzling yellow of a different world, the world of the Jurassic.

On this drive I took a several-hour break to see this transition. I went to a farm where Roger Smith had worked in the 1990s with James Kitching. The farmer and his wife remembered Roger and Kitch, and they welcomed me. I drove out behind their spacious house and climbed up into the hills. On a sparkling, perfect African day, I walked these outcrops and marveled at the treasures. Bones were there in profusion, and in the lowest of the Jurassic rocks, the bones spoke but one word—"dinosaur." This was new. And so, too, was my solitude. I had broken free of the Permian work. And I had broken free of Roger as well. It was time for both of us to move on.

Giant limb bones, broken vertebrae, the odd teeth of saurian herbivores—all were jutting from the flanks of the hill I had climbed. A hot sun baked all; to the north lay Kruger Park and the great herds of Africa. But here I was on the outcrop, a traveler to the past. Here were the bony remains of the Age of Dinosaurs, inheritors of the Gorgon's world. Hail to the kings. Their legions would hold sway for another hundred million years from the rocks I stood upon.

I knew then, as I know now, what the fate of the dinosaurs would be. A comet would end their rule in the now well-known K/T catastrophe. What I did not know was why there were dinosaurs at all. Or that the clue to the demise of their predecessors, the mammal-like reptiles of the Permian that I had so long studied, was in fact all around me, the most obvious clue of all. I also didn't know that dinosaurs, and their direct descendants the birds, are far more different from us mammals than anyone had supposed—and that these differences are direct results of the catastrophe that caused the P/T extinction. None of this was at all evident as I sat amid these latest Triassic rocks. Only the redness was evident. Sometimes the most obvious is the least evident. The redness of the rocks.

From the end of the Permian through the whole of the Triassic, until the Jurassic strata exposed here in this South African valley, all of the rocks showed a single unifying characteristic: All were red. And it was not just the South African rocks that were red: All over the world the Triassic and into the lower Jurassic is red. While there are other red beds at other times in Earth's history, this is the only period in the last 500 million years when one color dominates all rocks for such a long time interval. More than 50 million years of red. It is too obvious. Red. The color of rust. The color of oxidation. All around the world, rocks turned red as they oxidized. And in so doing this global rusting created two enormous biotic catastrophes: the Permian/Triassic extinction followed 50 million years later by the Triassic/Jurassic extinction. I now think I know what caused these extinctions.

On two occasions the oxygen in our atmosphere plunged to

very low levels as it became tied up in rocks that are now visible all over the world—levels so low, in fact, that any poor human finding him- or herself back in time from about 253 to about 175 million years ago would very quickly suffer from altitude sickness, even at sea level. And at higher altitudes, even as little as the mile-high elevation of Denver, death would ensue. This drop in oxygen pressure, combined with the temperature jumps at the end of the Permian (and again at the end of the Triassic), conspired to kill off much of the world's biota. It was not until the middle of 2003 that this would become clear to me. Now I am convinced that two things caused the Permian extinction: a sudden rise both in temperature and carbon dioxide near the end of the Permian Period and an equally sudden drop of oxygen in the atmosphere. Heat and asphyxiation, the two agents of the long mysterious mass extinction.

Coming to this conclusion involved a number of circuitous revelations. First, in late 2001, my friend Greg Retallack, fresh from our visit to the Karoo, presented a paper with a startling conclusion to the Geological Society of America: He proposed that *Lystrosaurus* survived the P/T extinction because it was preadapted for living at high altitude. It turns out that *Lystrosaurus*, virtually alone among the mammal-like reptiles, had an enormous rib cage and most probably large lungs within it. Was this evidence that *Lystrosaurus* first evolved at high altitude somewhere in the late Permian and then migrated into the lower elevation Karoo? I read this paper and shook my head. What was Greg thinking? But very quickly I began to understand his drift. Perhaps *Lystrosaurus had* evolved at high altitude, where oxygen is lower. Or perhaps *Lystrosaurus* evolved in a world where there was less oxygen not only at altitude, but everywhere.

We already knew that carbon dioxide levels skyrocketed at the end of the Permian, caused in part by sudden release of methane into the atmosphere and in part by the extrusion of carbon dioxide-rich volcanic gases produced by the Siberian traps flood basalts. We also believed that the time of the Permian/Triassic extinctions witnessed a sudden heating of the Earth by as much as 6 degrees C. Many workers, me included, had thought that the temperature rise would explain the mass extinctions, but we were all uneasy with this explanation. Could so many of the Earth's species have been so suddenly killed off by temperature increase alone?

Then in early 2003, I received a paper by Bob Berner. For the first time he showed that a rise in atmospheric carbon dioxide would create drop in the amount of oxygen in the atmosphere. He had graphs showing the critical interval of time. When I saw these I was astounded: Berner showed that the amount of oxygen in the atmosphere plummeted at the end of the Permian from levels somewhat higher that the 21 percent we enjoy today (almost all the rest of our atmosphere is nitrogen) to levels of 15 percent or less. At the P/T boundary and then again at the end of the Triassic, oxygen levels dropped to as low as 10 percent. Generally, such lowered oxygen levels are found only at altitudes above 15,000 feet, and at that altitude the entire pressure of the atmosphere lessens. During the Permian (and throughout the Triassic) the atmospheric pressure stayed the same, but the amount of oxygen dramatically dropped.

Here was a mechanism explaining the P/T. About a million years or so before the end of the Permian, ocean sea level began to drop. Organic-rich sediment, like that being preserved today in the Black Sea, was exposed to air. The organic material in this sediment had been deposited in an oxygen-free sea bottom. When exposed to air it started to oxidize—or rust. So much of this sediment was liberated that it began to pull oxygen molecules out of the air and convert them to sedimentary rock—the red rocks of the end of the Permian and the entirety of the Triassic—and the level of oxygen in the atmosphere dropped. At the same time, the Siberian trap volcanic fields began to form, and huge volumes of carbon dioxide gas rose into the sky. Carbon dioxide, as we know to our sorrow, is a powerful greenhouse gas. As it entered the atmosphere the whole world quickly started to warm. As a consequence, frozen methane on the continental edges and in the high latitudes thawed, liberating gaseous methane into the atmosphere. Methane is unstable. It rapidly converts to carbon dioxide, but does so only by taking more oxygen out of the atmosphere, for the conversion of methane to carbon dioxide takes oxygen molecules. The world got suddenly hotter, and drier, and all of its creatures— plants, animals, those in the sea, those on land—began to asphyxiate from lack of oxygen and too much carbon dioxide. Only a few survived. If humans had been there we might not have made it.

In the end, it now seems that the Gorgon died out horribly. It died of asphyxiation, almost taking the dreams of a distant hominid intelligence with it.

Of course, as with all research, the implications of this new information is far reaching and leads to entirely new directions in research. It's my hope that it will eventually uncover information about the evolution of humans that we've never even imagined.

In early 2003, I was in the lab of Joe Kirschvink in Pasadena when one of his students told me that it had recently been discovered that dinosaurs had the same lungs as birds. By chance, I had just been puzzling over bird lungs. I had learned that birds had been observed flying high over Mt. Everest, at more than 30,000 feet. No mammal could live at such altitude. Birds have a peculiar lung system with attached air sacs that make them very efficient at oxygen acquisition. These sacs are throughout their body, and even permeate their bones. The prevailing wisdom is that birds need this type of lung for flying. Yet the new information indicated that dinosaurs, like birds, had air sacs as well as lungs. Why would a 20-ton dinosaur need such a lung? A light bulb went on. I looked at Joe with delight as I mentally put together the Berner oxygen findings, Greg Retallack's idea about *Lystrosaurus*, and all the red rocks I had seen on my last drive out of the Karoo. What if birds had evolved not for flight, but to allow survival in a world with low oxygen?

According to Bob Berner, the world did not return to the oxygen levels of today until after the evolution of the first true birds in the middle Jurassic, well into the so-called Age of Dinosaurs. The ancestral dinosaurs—and their weird, bird-like lungs—first evolved in the Triassic *at a time of radically low oxygen levels*. What if dinosaurs, like their Jurassic descendants, the birds, were in fact new evolutionary forms adapted to thrive in low oxygen?

We mammals came from the mammal-like reptiles, which evolved at a time in Earth history when there was high oxygen, and we have a lung type that is very poor in low-oxygen settings. In fact, the only creatures worse than us are the very primitive lizards—the same types of primitive reptiles that also evolved when oxygen in the atmosphere was high, even higher than today, back about 300 million years ago in the Coal Age, or Carboniferous Period. Our ancestors did not need particularly good lungs, at first, in their oxygen-rich Paleozoic world. But to thrive in the Triassic a new type of lung was needed.

I am proposing that the dinosaurs and their descendants, the true birds, came about as a result of low oxygen. But perhaps there is even more that owes its existence to this strange time in Earth history. Did live birth in mammals result from this time? Did warm bloodedness? This idea is novel and proposed for the first time in this book.

All the pieces seemed to fit. There were two times of very low oxygen—at the end of the Permian, 250 million years ago, and at the end of the Triassic, 200 million years ago. Both times coincide with two of the most profound mass extinctions in Earth history. We have seen how devastating the Permian extinction was. At the end of the Triassic, the extinctions were again devastating for mammals and mammal-like reptiles, but did almost nothing to the dinosaurs. By 200 million years ago their kind was supremely well adapted for the low oxygen.

The time of low oxygen ended in the middle Jurassic. The rocks turned yellow again. Mammals could once again compete, yet the dinosaurs held sway until killed off by another chance event—the comet at the end of the Cretaceous. But that's another story.

Legacy and Lessons
of a Catastrophe

Are We Living on a Safe Planet?

From 1980 until the first years of the new century, the discipline of paleontology undertook a radical trajectory. Up through the late 1970s, mass extinctions were thought to have been rather gradual events of multimillion-year duration that caused a slow change in the flora and fauna of the earth. As species gradually disappeared, others slowly took their place. The word "catastrophe" was anathema in the geological sciences, and the mass extinctions were never characterized as catastrophes. Besides, there were not very many of them. And then, in 1980, the Alvarez hypothesis that the dinosaurs (and much else) went extinct due to the sudden catastrophic environmental effects following an asteroid impact on the earth caused a paradigm shift—rapid mass extinctions can (and have been) caused by the impacts of extraterrestrial bodies.

The paleontological community was mobilized into action by this new paradigm, causing many investigators to change their research programs and begin active pursuits of extinction questions. For if the dinosaurs were killed off by the impact of a relatively large celestial body with the earth, surely others of the mass extinctions might have a common cause. Not surprisingly, it was soon posited that many, perhaps most, of the other mass extinctions had been caused by large-body impact.

Yet not only was there new understanding about the rapidity and cause of the mass extinctions, but at the same time an important change occurred as well about the evolutionary consequences of mass extinctions. In the aftermath of the Alvarez revolution, two new views gained currency and have attained the level of dogma two decades later:

First, that mass extinctions, because of their lethality, reset the evolutionary agenda prevailing at the time. With the disappearance of so many species in so short a time (including forms that might be called evolutionary incumbents, dominant groups that held back the evolution of other, nondominant groups), entirely new body plans could and would arise to meet the opportunity of a suddenly low-diversity biota. *Mass extinctions are thus agents of evolutionary novelty.* Second, that mass extinctions, although followed by long periods of low diversity (as the world recovered and species evolved to take the place of the newly dead), eventually paved the way for an even higher biodiversity than was present prior to the catastrophe. *Mass extinctions were thus agents creating higher levels of biodiversity. In fact, some minimum number might be necessary to populate and then keep a planet stocked with higher organisms.*

Let me blaspheme. I suspect neither of these now hoary "truths" to be the real gospel. I now believe that the Permian extinction yields but a single important lesson: Planets with higher life—those with the equivalents of our animals and plants—*can be rendered abiotic*, and that asteroid impact can certainly do the job. Let me be more explicit. Mass extinctions reset the composition of animals and plants on a world, but in the long run this may matter little. In his book *Crucible of Creation*, the British paleontologist Simon Conway Morris attacked Stephen Jay Gould's suggestions that chance survivors of the Burgess Shale fauna determined the biological makeup of the present-day animal fauna. Conway Morris took the opposite tack—that even if *Pikaia*, the ancestral vertebrate, had been wiped out, there would still be vertebrates on the planet today, because the vertebrate body plan is too good a structure for aquatic and terrestrial use not to reevolve.

Does this mean that importance of mass extinctions has been overestimated? Not at all. We have learned several invaluable lessons: Impacts can cause mass extinctions, and mass extinctions have the potential of wiping out entire kingdoms of organisms, most readily the multicellular animals and plants. Even more important, I believe that the number and temporal spacing of mass extinctions is the most important aspect in assessing their effect on a biota. For instance, if there had been even a few more P/T and K/T events, at closer spacing, then we might indeed live upon a very impoverished world, or one where the Age of Bacteria had returned following the complete extermination of animal life. In my view the message is that, at least on earth, we have seen only failed killers, and there have been few of them. Yet we now know, through these earlier botched assassination attempts, that real celestial killers are at large and that the Permian and Cretaceous extinctions could certainly have spelled the end of animal life on the planet had the falling rocks been even only slightly larger in size—or had there perhaps been ten times as many of them in the last 500 million years (the time we might call the Age of Animals). In other words, old age may not be the only way that a living planet dies. As with Caesar, one, or even several, knife blows may not be enough to kill the beasts.

But that is getting a bit ahead of ourselves. Lets us return to 1980 and the heady days soon after the announcement by the Alvarez group.

The sea change resulting from the celebrated Alvarez theory was that the Principle of Catastrophism, long ago discredited by Hutton and Lyell, was alive (or dead, forgive the pun) and well. Yet a subtler, and still little appreciated, change in thinking occurred as well. Because the Alvarez hypothesis involved an element from outer space, specialists on comets, asteroids, and even on the nature and structure of plants, stars, and galaxies entered into the fevered discussions and studies. For the first time, paleontologists and geologists, the longtime keepers of stratigraphic lore, began to attend scientific meetings with several entirely new fraternities: astronomers, planetary geologists, and physicists. The concept that life on earth was affected by the history of non-earth bodies was of interest to all these groups. The various scientists, stimulated by these new associations, began to think in new ways, to consider new types of questions. For the first time, the earth began to be viewed as one known case of history among the many planets certainly existing in our galaxy, and in the universe as a whole. Was our particular history—of long-term biological evolution and diversification punctuated by (and perhaps abetted by) occasional mass extinctions—*typical* of populated planets? A consequence of this new view was the realization that planets themselves might undergo variable amounts of risk. This point was articulated by the University of Chicago paleontologist David Raup in his 1991 book, *Extinction: Bad Genes or Bad Luck?* when he posed the provocative question, Are we living on a safe planet?

To Raup the answer seemed to be no.

One idea I think most of us share is that the earth is a pretty safe and benevolent place to live. . . . Is all of this true or merely a

fairy tale to comfort us? Is there more to it? I think there is. Almost all species in the past failed. If they died out gradually and
quietly and if they deserved to die because of some inferiority,
then our good feelings about earth can remain intact. But if they
died violently and without having done anything wrong, then
our planet may not be such a safe place.

The next logical point—*Safe* relative to what?—was not addressed until the early years of the twenty-first century. It was
part of a new scientific movement that, in my opinion, was
largely an offshoot of the multidisciplinary research efforts by
countless scientists in the wake of the Alvarez impact hypothesis.
This new scientific discipline is now known as astrobiology.

Astrobiology is the study of how and where life might occur
beyond earth. Yet it also is concerned with where and why life
might not be present and how life might die off on a given planet.
Perhaps life is common in the universe, but has difficulty surviving for long periods because of a heavier rain of comets and asteroids than has afflicted the earth. On how many planets has life
developed, only to be snuffed out by some slightly more catastrophic equivalent of the Permian or Cretaceous extinctions?
And even on our own planet, if the K/T asteroid hit the earth
during the first million years after the Permian-extinction event,
would any animal have survived in an already stressed and impoverished world? Could the Gorgon, my metaphor for the great
Permian extinction, really be the giant comets found throughout
space, raining into potentially inhabitable planets with murder-
ous frequency?

Yet if space is as dangerous as most astrobiologists think it is,
why is it that we have survived for so long on this planet? After
all, our solar system is filled with, and surrounded by, huge in-
ventories of comets. How safe are we on this earth from another

such catastrophe? Will the next millennium—or the next million years, or even the next billion years—witness the impact of a comet large enough to annihilate life on earth? A great Hollywood scenario, but, I've come to believe, not so very likely. Somehow, in the long years of Karoo research, I have concluded that the mass extinctions on earth have been minor aberrations on a fantastically safe—or perhaps just lucky—planet.

The greatest surprise was not that the extinction was so sudden, or that it so changed the world, though both are true. The greatest surprise is that phenomena such as the Permian extinction *have been so rare*.

Over the past decade, a new truth seems evident: Instead of finding ever more mass extinctions, of greater severity, we are finding fewer. The so-called Big Five—the mass extinctions of the Ordovician, Devonian, Permian, Triassic, and Cretaceous—have been whittled down to a Big Three. New work by Mike Foote and his students at the University of Chicago suggests that the first two of these events, while undoubtedly major crises of some sort, were not in the class of the later Mesozoic events. My own new work on another of the Big Five, the extinction at the end of the Triassic period, also suggests that it was never a threat to end animal life on the planet. In the 500 million years of animal life on the planet, the most consequential extinction of all, at the end of the Permian, wreaked enormous and undoubted extinctions but came nowhere near to eliminating animal and planet life. For all that wreckage, within 10 million years the world had caught up to its prior biodiversity and then surpassed it.

The 1980 discovery that a mass extinction had been caused by an asteroid impact was revolutionary. It got people thinking about life on other worlds as well as about life on this one. One of the questions that came to mind involved the frequency of asteroid and comet impacts on the earth. By examining the size and frequency of meteor-impact craters, Gene Shoemaker and others, in the early 1990s, calculated that we might expect an impact by an asteroid or comet of about the same size as that causing the K/T extinction every 100 million years. This frequency might even roughly fit the facts on earth: There have been five major extinctions in the past 500 million years (even though the K/T is still the only one undoubtedly caused by impact). But a salient fact remains—that the K/T asteroid came nowhere near wiping out all animals and plant species. We took this best shot, reeled a bit, and got back to business relatively quickly.

Yet might there be other places in the galaxy where that frequency and/or the size of the rocks raining in would be higher or lower. What factors might influence impact rate and size? And, supposing that impact rate varied according to the celestial neighborhood, might this not translate to variable rates of mass extinctions, depending on where in the galaxy a particular set of stars and planets resided? Perhaps there are safe and less safe celestial neighborhoods, places where the Gorgon is barred entry, others where it constantly lives.

I went to Africa in the early 1990s, to a new country, to start a new scientific journey of discovery. I went with the mission of that time: Was the Permian extinction, like the Cretaceous extinction, caused by a large-body impact with the earth? And if two or more such extinctions were brought about by celestial killers, how long until the next, potentially planet-killing, impact might occur? Africa seemed to hold the key.

I went for so many reasons. To visit a new place, to study a new problem. Corny as it may sound, I also went to find adventure, and perhaps fame, in making a great discovery. What scientist does *not* dream of the great discovery, the moment of clarity when a momentous problem is solved? And what scientist is not human enough to wish for the benefits that can accrue from such discoveries? Was I much more altruistic? It's hard to remember motives when more than a decade has passed. Most of all I went to lend my two cents' worth in figuring out What Did It. What pulled off such an awful extinction, 250 million years ago? I went for curiosity. Learning the identity of the perpetrator of that awful crime seemed the most important thing.

A decade passed. A fifth of my life. And in the end change, that winner of all battles, took its due. The Karoo Paleontology Unit was broken up. The first to go was Georgie. She was dogged by South African immigration and had to leave the country every few months to get her visa renewed. She was refused permanent resident status and could no longer join the fossil hunts that so fulfilled her. Soon after, Paul October quit the team. He was transferred in the museum to a menial job in the planetarium. Only Hedi Stummer still remains.

The South African Museum was given a new director, a political appointee, and the entire scientific mission of the museum was reexamined in light of the economic priorities of South Africa. All curatorial staff was required to resign, and a commission was assembled to decide if they would be rehired. Roger Smith went through this traumatic and insensitive ordeal. He got his job back, but enormous bitterness was engendered. He began to spend ever more time at a small museum field station several hours north of Cape Town, a place where Ice Age mammals could be found in abundance. With this change he spends less and less time in the Karoo. He still runs his ultramarathons and keeps more to himself. I receive the occasional e-mail, but our time together seems to have passed. (Though he does have me to thank for a last wonderful adventure—he'll be accompanying Greg Retallack and Luann Becker to Antarctica to look at the Permian rocks there. When Luann learned that Roger had invited me she demanded that the invitation be rescinded, so Roger will go in my stead.)

South Africa itself remains in a storm of change. It is too soon to tell if the country will pass through this storm and emerge into a truly vibrant multiracial country—the economic engine of the African continent—or if it will follow the model of nearby Zimbabwe and become a racist enclave and dictatorship where crime and murder are the laws of the land.

As for me, I am involved in a new adventure, the story of the rise and fall of oxygen on our planet and how that shaped us, perhaps most importantly in a great mass extinction of 250 million years ago. But mostly I am staying home and watching over my son as he grows up. I'm a father who no longer needs to travel to far African deserts to discover what can be found in a boy's smile and a wife's calming embrace.

References

Alvarez, L., et al. 1980, Extra-terrestrial cause for the Cretaceous-Tertiary extinction. *Science* 208: 1094–108.

Anderson, J., and A. Cruickshank. 1978. The biostratigraphy of the Permian and the Triassic: a review of the classification and distribution of Permo-Triassic tetrapods. *Palaeont. Afr.* 21: 15–44.

Benton, M. 1995. Diversification and extinction in the history of life. *Science* 268:52–58.

Broom, R. 1932. *The Mammal-like Reptiles of South Africa and the Origin of Mammals.* London: H. Witherby.

Brown, L., C. Flavin, and H. French. 1999. *State of the World, 1999.* New York: Norton/Worldwatch Books.

Caldeira, K., and J. Kasting. 1992. The life span of the biosphere revisited. *Nature* 360:721–723.

Cohen, J. 1995. *How Many People Can the Earth Support?* New York: Norton.

Covey, C., et al. 1994. Global climatic effects of atmospheric dust from an asteroid or comet impact on Earth. *Global and Planetary Change* 9:263–273.

Ehrlich, P. 1987. Population biology, conservation biology, and the future of humanity. *Bioscience* 37:757–763.

Ellis, J., and D. Schramm. 1995, Could a supernova explosion have caused a mass extinction? *Proc. Nat. Acad. Sci.* 92:235–238.

Erwin, D. 1993. *The Great Paleozoic Crisis; Life and Death in the Permian.* New York: Columbia University Press.

———. 1994. The Permo-Triassic extinction. *Nature* 367: 231–36.

Gehrels, T., ed. 1994. *Hazards Due to Comets and Asteroids.* University of Arizona Press.

Goudie, A., and H. Viles. 1997. *The Earth Transformed.* Blackwell Publications.

Gribbin, J. 1990. *Hothouse Earth.* New York: Grove Weidenfeld.

Grotzinger, J. P., et al. 1995. New biostratigraphic and geochronological constraints on early animal evolution. *Science* 270:598–604.

Hallam, A. 1994. The earliest Triassic as an anoxic event, and its relationship to the End-Paleozoic mass extinction. *Canadian Society of Petroleum Geologists* mem. 17:797–804.

———, and P. Wignall. 1997. *Mass Extinctions and Their Aftermath.* Oxford: Oxford Univ. Press.

Holser, W., et al. 1989. A unique geochemical record at the Permian Triassic boundary. *Nature* 337:39–44.

Hotton, N. 1967. Stratigraphy and sedimentation in the Beaufort Series (Permian-Triassic), South Africa. Essays in Paleontology and Stratigraphy. *University of Kansas* special publication. 2:390–428.

Hsu, K., and J. Mckenzie. 1990. Carbon isotope anomalies at era boundaries; global catastrophes and their ultimate cause. *Geol. Soc. Am.* special paper 247: 61–70.

Isozaki, Y. 1994. Superanoxia across the Permo-Triassic boundary record in accreted deep-sea pelagic chert in Japan, in Global Environments and Resources. *Canadian Society of Petroleum Geologists* mem. 17: pp. 805–12.

Jablonski, D. 1991. Extinctions: a paleontological perspective. *Science* 253: 754–757.

Keyser, A., and R. Smith. 1979. Vertebrate biozonation of the Beaufort Group with special reference to the Western Karoo Basin. *Mem. Geol. Surv. S. Africa* 12: 1–36.

King, G. 1990. Dicynodonts and the end Permian event. *Palaeontologia africana* 27: 31–39.

Kirchner, J.W., and A. Weil. 2000. Delayed biological recovery from extinctions throughout the fossil record. *Nature* 404: 177–180.

Kirschvink, J. 1992. A Paleogeographic Model for Vendian and Cambrian Time. Chapter XII in:, in: *The Proterozoic Biosphere: A Multidisciplinary Study.,* J.W. Schopf, C. Klein and D. Des Maris, eds., 567–81, Cambridge: Cambridge University Press.

Kitching, J. 1977. Distribution of the Karoo vertebrate fauna, *Price Inst. Pal. Res.* mem. B.:1–131.

Knoll, A., et al. 1996. Comparative Earth history and Late Permian mass extinction. *Science* 273: 452–57.

Kruess, A. and T. Tscharntke. 1994. Habitat fragmentation, species loss and biological control. *Science* 264:1581–84.

Lovelock, J. 1979. *Gaia, a New Look at Life on Earth.* Oxford: Oxford University Press.

McKinney, M., ed. 1998. *Diversity Dynamics.* New York: Columbia University Press.

MacLeod, K., et al. 1997. Stable isotope results from the Permian extinction boundary, Karoo basin, South Africa. Geological Society of America, annual meetings, abstracts with programs.

Maher, K. A., and J. D. Stevenson. 1988. Impact frustration of the origin of life. *Nature* 331: 612–14.

Marshall, C., and P. Ward. 1996. Sudden and gradual molluscan extinctions in the latest Cretaceous of Western European Tethys. *Science* 274: 1360–63.

May, R. 1988. How many species are there on Earth? *Science* 241: 1441–49.

Morante, R. 1996. Permian and early Triassic isotopic records of carbon and strontium events in Australia and a scenario of events about the Permian-Triassic boundary. *Historical Geology* 11: 289–310.

Myers, N. 1985. The end of the lines. *Natural History* 94:2–12.

———. 1993. Questions of mass extinction. *Biodiversity and Conservation* 2:2–17.

———. 1996. The biodiversity crisis and the future of evolution. *The Environmentalist* 16: 124–36.

Paine, R., M. Tegner, and E. Johnson. 1998. Compounded Perturbations Yield Ecological Surprises. *Ecosystem* 1: 535–45.

Pimm, S. 1991. *The Balance of Nature: Ecological Issues in the Conservation of Species and Communities.* Chicago: University of Chicago Press.

Pimm, S., et al. 1995. The future of biodiversity. *Science* 269:347–54.

Pope, K., et al. 1994. Impact winter and the Cretaceous Tertiary extinctions: results of a Chicxulub asteroid impact model. *Earth and Planetary Science Express* 128:719–25.

Rampino, M., and K. Caldeira. 1993. Major episodes of geologic change: correlation, time structure and possible causes: *Earth Planet. Sci. Lett.* 114:215–27.

Raup, D. 1979. Size of the Permo-Triassic bottleneck and its evolutionary implications. *Science* 206:217–18.

———. 1990. Impact as a general cause of extinction: a feasibility test, in *Global Catastrophes in Earth History,* V. Sharpton and P. Ward, eds. *Geol. Soc. Am.* special paper 247: 27–32.

————. 1991. A kill curve for Phanerozoic marine species. *Paleobiology* 17:37–48.

————, and J. Sepkoski. 1984. Periodicity of extinction in the geologic past. Proceedings of the National Academy of Sciences, A81, pp. 801–5.

Retallack, G. 1995. Permian-Triassic crisis on land. *Science* 267:77–80.

————. 1999. Postapocalyptic greenhouse paleocliamate revealed by Earliest Triassic paleosols in the Sandhy Basin, Australia. *GSA Bulletin* 111: 52–70.

————, and E. Krull. 1999. Landscape ecological shift at the Permian/Triassic boundary in Antarctica. *Australian Journal of Earth Sciences* 46: 785–812.

Sepkoski, J. 1984. A model of Phanerozoic taxonomic diversity. *Paleobiology* 10:246–67.

Rubidge, B. 1995. Biostratigraphy of the Beaufort Group (Karoo Sequence), South Africa. Geological Survey of South Africa, Biostratigraphic Series, no.1, pp. 1–43.

Schindewolf, O. 1963. Neokatastrophismus? *Zeitung der deutschen geologische Gesellschaften* 114:430–45.

Schwartzman, D., M. McMenamin, and T. Volk. 1993. Did surface temperatures constrain microbial evolution? *BioScience* 43:390–93.

Sheehan, P., et al. 1991. Sudden extinction of the dinosaurs: Latest Cretaceous, Upper Great Plains, U.S.A. *Science* 254: 835–39.

Sleep, N. H., et al. 1989. Annihilation of ecosystems by large asteroid impacts on the Earth. *Nature* 342: 139–42.

Sigurdsson, H., S. D'hondt, and S. Carey. 1992. The impact of the Cretaceous-Tertiary bolide on evaporite terrain and generation of major sulfuric acid aerosol. *Earth Planetary Science* letters 109: 543–59.

Smith, R. 1990. A review of stratigraphy and sedimentary environments of the Karoo Basin of South Africa. *Journal of African Earth Sciences* 10:117–37.

————. 1995. Changing fluvial environments across the Permian-Triassic boundary in the Karoo Basin, South Africa and possible causes of tetrapod extinctions. *Paleo.Paleo. Paleo.* 117:81–104.

Stanley, S. 1987. *Extinctions.* W. H. Freeman.

————, and X. Yang. 1994. A double mass extinction at the end of the Paleozoic Era. *Science* 266:1340–44.

Stuart, C. and T. Stuart. A field guide to the larger animals of Africa. Struik Publishers: Cape Town. 390 p.

Teichert, C. 1990. The end-Permian Extinction. In Kauffman, E. and O. Walliser, eds. *Global Events in Earth History:* 161–190.

Thackeray, J., N. van der Merwe, J. Lee-thorpe, A. Sillen, J. Lanham, R.Smith, A. Keyser, P. Montiero. 1990. Changes in carbon isotope ratios in the late Permian recorded in therapsid tooth apatite. *Nature* 347:751–753.

Valentine, J. W. 1994. Late Precambrian bilaterans: grades and clades. *Proceedings of the National Academy of Sciences* 91:6751–6757.

———, D. H. Erwin, and D. Jablonski. 1996. Developmental evolution of metazoan body plans: The fossil evidence. *Developmental Biology* 173:373–381.

Ward, P., 1990, The Cretaceous/Tertiary extinctions in the marine realm; a 1990 perspective: *Geol. Soc. Am.* special paper 247: 425–432.

———. 1994. *The End of Evolution.* New York: Bantam Doubleday Dell.

———. 1990. A review of Maastrichtian ammonite ranges. *Geol. Soc. Am.* special paper 247:519–30.

———, and D. Brownlee. 2000. *Rare Earth.* Copernicus (Springer Verlag).

———, et al. 1991. Ammonite and inoceramid bivalve extinction patterns in Cretaceous-Tertiary boundary sections of the Biscay Region (southwest France, northern Spain). *Geology* 19:1181–84

———, and W. Kennedy. 1993. Maastrichtian ammonites from the Biscay region (France and Spain). *Journal of Paleontology* mem. 34:67:58.

Index

FOR THE BEST IN PAPERBACKS, LOOK FOR THE ⟨🐧⟩